ACID RAIN RESEARCH FOCUS

ACID RAIN RESEARCH FOCUS

LYMAN G. ROGLESFIELD
EDITOR

Nova Science Publishers, Inc.
New York

For permission to use material from this book please contact us:
Telephone 631-231-7269; Fax 631-231-8175
Web Site: http://www.novapublishers.com

LIBRARY OF CONGRESS CATALOGING-IN-PUBLICATION DATA

Acid rain research focus / Lyman G. Roglesfield, Ed.
p. cm.
ISBN 978-1-60456-373-3 (hardcover)
1. Acid rain--Environmental aspects--Research. I. Roglesfield, Lyman G.
TD196.A25A32 2008
628.5'32--dc22 2008001096

Published by Nova Science Publishers, Inc. ✦ New York

CONTENTS

PREFACE

Acid rain is rain or any other form of precipitation which is unusually acidic. It has harmful effects on plants, aquatic animals and buildings. Acid rain is mostly caused by human emissions of sulfur and nitrogen compounds which react in the atmosphere to produce acids. In recent years, many governments have introduced laws to reduce these emissions. This new book examines new research results from around the world.

Chapter 1 can be divided into three main parts. The first one summarizes on the one hand a method for determining the acid dew point temperature (ADT) over a regenerator in a 370 MW reference thermal power plant for two types of fuel, such as: *"fuel N^{er} 2"* and *"low sulphur fuel"* respectively and on the other hand, it shows how a changeover to this latter fuel would increase the useful lifetime of some of those equipments directly involved such are heat exchangers, reducing substantially the cost of maintenance due to the considerable decreasing of the surface and time exposed to acid depositions with the subsequent increasing in boiler's performance and reliability.

The second part is focused on new studies on the effect that gradual replacement of "*fuel N^{er} 2*" by "*low sulphur fuel*" would have with regard to the environment. Thus, an analytical comparison (in terms of emissions) was carried out between both types of fuel-fired power plant schemes, showing this way the important advantages of the last ones, especially in terms of emissions and efficiency. Finally, a comparison with experimental values available from the gas analyzer has also been performed, showing a very good agreement with the data previously obtained in the calculations carried out here, so in this way that analytical procedure could be considered as validated. The authors will also evaluate the re-turning fuel oil boilers into natural gas ones which has risen over the last years. In this way, conventional fuel oil fired power plants are gradually replaced by natural gas-fired combined cycles, reducing sensitively the CO_2 emissions while SO_2 emissions are almost negligible. European Directives will be summarised showing the ceiling emissions.

Finally, the later study addresses the experimental optimization of a specific operation mode recently carried out over a thermal power plant working at minimum technical load (130 MW) by using a profitable combination of low-sulphur fuel burners and natural gas igniters, traditionally used as combustion support, analysing the economic and environmental advantages, considering also the fuels prices fluctuations in the context of the global market.

As explained in Chapter 2, acid precipitation represents the deposition of acidic components in rain, snow, dew, or dry particles. Acid rain occurs when sulfur dioxide and nitrogen oxides are emitted into the atmosphere, undergo chemical transformations, and are absorbed by water droplets in clouds. The droplets fall to Earth and increase the acidity of the soil, lakes and streams. Acid rain can have adverse impacts on forests, freshwaters and soils, insects, aquatic life forms, and can cause damage to buildings. In absence of precipitation, acid deposition occurs via dry deposition, a significant process, especially in dryer climates, or during periods of drought. Wet deposition of acids occurs when precipitation removes acids from the atmosphere and delivers it to the Earth's surface. This can result from the deposition of acids produced in the raindrops or by the precipitation removing the aerosol particles either in-cloud or below-cloud. In this chapter, the authors address in detail the wet removal of aerosol particles by rain. Wet removal of atmospheric aerosols by falling precipitation is a major removal mechanism for airborne particles. In-cloud wet removal is the dominant mechanism and the rate of aerosol removal in cloud is largely dependent on rainfall rate and the collision efficiency between raindrops and aerosol particles incorporated in cloud droplets. Below-cloud removal rate of aerosols by rainfall is dependent on the rainfall rate, collision efficiency between raindrops and aerosol particles, and on aerosol size distribution. Results presented show the link between the aerosol removal rate and the microphysical characteristics of precipitation and aerosol such as raindrop size distribution, aerosol size spectrum, collision mechanisms, and environmental variables.

Urban areas generally undergo high pollution loads, specially due to traffic and industrial plants, and particular microclimate conditions, depending on wind pattern and topographic structure. This results in higher contamination than in natural sites, with harmful and undesired effects on human health, ecosystem and materials, in particular those constituting cultural heritage. Sampling and analysis of wet and dry atmospheric deposition can provide a clear picture of the environmental quality of an ecosystem and can lead to actions aimed at reducing pollutant emissions in the atmosphere. Chapter 3 describes the results of a nine years monitoring (1995-2003) of wet and dry depositions collected in the urban site of Bologna, with a DDAS (Dry Deposition on Aquatic Surface) sampler. Bologna is located in the Po Valley (Northern Italy), a heavily industrialized and densely populated region, characterized by a typical atmospheric circulation frequently associated with the air mass stagnation conditions.

The purposes of this study were to investigate the chemical composition of both wet and dry depositions, with regard to the main cations (H^+, Ca^{2+}, Mg^{2+}, Na^+, K^+, NH_4^+), anions (SO_4^{2-}, NO_3^-, Cl^-) and heavy metals (Cd, Cr, Cu, Ni, Pb, V, Zn), to examine the temporal trends of the main ions, to identify the contribution of different sources to the chemical composition, in particular distinguishing between natural and anthropogenic ones, and to point out local, regional and long-range influences. The study of correlation between ion pairs and data factor analysis represented a useful tool to this aim.

Three main factors were identified which affected chemical composition in the 9 years survey: anthropogenic, marine and terrigenous contributions, acting respectively on a local, regional and both local and long-range scale. The separate study of wet and dry deposition revealed a predominance of anthropogenic species fallout in the former mechanism, due to their presence in the finest particles, compared to those from natural sources. Finally, a clear decrease of free acidity in atmospheric deposition was also observed, associated to the lowering of anthropogenic species, as nitrate and sulphate ions; this suggests an improvement of air quality in Bologna urban area, due to the recent advances in pollutant control.

Acid rain damages many kinds of plants and living things by reaching cations from soil and lake water, and has become serious problems in the world. To protect soil and lake water from acidification by acidic rain, calcium carbonate powder has been used for neutralization of acid in Europe and Canada but they cause another problem. Thus, the authors reviewed novel bioremediation methods of acidic soil and water with microbes in Chapter 4. They screened microorganisms showing a high ability to neutralize acids. These strains of fungi that were isolated from soil microorganisms showed high resistance to acid and high ability of neutralization. The fungi could grow in acidic media (pH 4.0) containing several kinds of acids and increased pH up to around 8.0. When they grew in LB medium (pH4.0) adjusted with nitric acid, NH_4^+ in the medium increased and NO_3^- decreased. Therefore, it is considered that the neutralization of nitric acid was caused by both a secretion of NH_4^+ and by the assimilation of NO_3^-.

In addition, the authors proposed a self-immobilization method (BSIS), in which a *Bacillus* strain forming aggregates and a strain bioremediating pollutants were co-immobilized in a shallow layer in soil. Bioremediation in the shallow layer would be effective for protection of ecosystems in soil. By the BSIS method, fungus cells could be co-immobilized with *Bacillus subtilis* in a shallow layer of model soil. The immobilized cells could successfully neutralize acidic solutions flowing through packed soil. This means that BSIS is applicable to neutralization of acid rain caused by SOx and NOx for relatively long time. The authors also applied a fungus *Rhizopus* strain to bioremediation of acidified water. The strain could grow and neutralize acid in pond water. To neutralize a large amount of water in a short period, the strain was immobilized in a cellulose tube. Although the volume ratio of the cell suspension to pond water was only one-fortieth, the pH value increased from 4.0 to 7.8 within 1 hour, and nitrates decreased below 0.02 g/l. These results suggest that the immobilized *R. delemar* cells could continuously neutralize pond water and this immobilization method is useful in neutralizing acidified water for a long period.

As presented in Chapter 5, acid rain's negative impact on world heritage buildings and human health in Vietnam is not very well known. Historical monuments, statues, and buildings are covered by black soot and fly ash, caused by fossil fuels in acid rain in Hue, Cam Phuc, and Cam Chinh in Central Vietnam. The sooty surfaces of study subjects show damage by not only acid rain, but also by fungi, lichens, and bacteria. Rainwater collected in January 2006 and August 2007 in 3 villages and observed by electron microscopy showed abundant carbon spherules and fly ash, associated with clay minerals, iron materials and microorganisms. Most Vietnamese drink well water and/or collected rainwater, especially in the countryside. Sometimes acid rain water shows a pH of 6-8, because of the effects of sea salts, but its electrical conductivity (EC) remains quite high (eg:300-1200 μ S/cm), suggesting high concentrations of SOx and NOx components. Chemical, biological, and environmental studies were performed on world heritage buildings in Dong Ha, whereas the rain and drinking water collected in Hanoi, Cam Chinh (30 homes) and Cam Phuc (32 homes) were studied, in association with concurrent health check interviews. Some heavy metals (Cr, Fe, Zn, Br, and Sr) and colon bacilla were found in the drinking water even after filtrations at Cam Phuc Village. In Cam Phuc, incidence of diseases such as anemia, roundworm, dermatitis, and symptoms of fatigue, loss of appetite, skin problems, stomachache, and dizziness is worse than that in Cam Chinh, indicating no correlation between former Agent Orange dispersion and current health conditions covered by this study. The worst health conditions were caused by drinking water in Cam Phuc village, a control area with no history of Agent Orange contamination. Further more, volatile organic compound concentrations of formaldehyde, acetaldehyde, acetone, propionaldehyde, hexaldehyde, and 2,5-

dimethylbenzaldehyd in both Cam Chinh and Cam Phuc villages were quite high. Especially high acetone concentrations (13-59 μ g/m^3) were detected both indoors and outdoors, and showed more than 100 times the acetone levels of the control standard in Japan. The origin of the acetone is unknown.

As explained in Chapter 6, the subtropical alpine forest ecosystem is vulnerable to acid rain deposited through the forest soils. The average sulfate and nitrate deposited by acid rain as precipitation amount to 50 kg SO_4^{2-} and 15-50 kg NO_3^- ha^{-1} year in Taiwan, respectively. Soil samples were leached with simulated acid rain solutions of various acidity and sulfate concentrations to evaluate their retention of cations and sulfates, testing and leaching with break through curves with simulated acid rain solutions. The pH buffer capacity of forest soils along an elevation sequence in central Taiwan was also investigated through conventional titration and mathematical calculation of proton affinity methods. The amounts of sorbed-SO_4^{2-} of two pedons extracted with NaH_2PO_4 (0.016 M) solutions in the range 0.14-1.09 mmol kg^{-1} soil were higher than that of native water-soluble sulfate (i.e., extracted with H_2O), in the range 0.15-0.41 mmol kg^{-1} soil. The sorption of sulfate did not fit well the Langmuir equation. The sulfate sorption capacity and the amount of released Al correlated well with increasing acidity of simulated acid rain added. Released K, Mg and Ca showed two periods of higher leaching pattern, having a trend similar to that the electric conductivity (EC) pattern. The pH buffer capacity (pHBCs) showed no significant differences from the titration curves of various pedons and horizons. However, mathematical calculation of proton affinity potentials showed the highest correlation at the lower elevation site with soil pH ($r = 0.90^{**}$), cation-exchange capacity (CEC) ($r = 0.91^{**}$), base saturation ($r = 0.97^{**}$), and X-ray non-crystalline sesquioxides ($r = 0.68^{**}$). It is important to evaluate the alpine forest soil pH buffer capacity by combining its proton consumption potentials.

In: Acid Rain Research Focus
Editor: Lyman G. Roglesfield, pp. 1-44

ISBN: 978-1-60456-373-3

Chapter 1

ENVIRONMENTAL AND ECONOMICAL COST OF THE ACID EMISSIONS INVOLVING THERMAL POWER PLANTS BURNING DIFFERENT FUELS, BASED ON AVAILABLE DATA

J.M. Blanco **[*]** ***and F. Peña***

Dpto. Máquinas y Motores Térmicos, Escuela Técnica Superior de Ingeniería, Universidad del País Vasco / E.H.U. Alameda de Urquijo s/n (48013) Bilbao. España

Abstract

This work can be divided into three main parts. The first one summarizes on the one hand a method for determining the acid dew point temperature (ADT) over a regenerator in a 370 MW reference thermal power plant for two types of fuel, such as: *"fuel N^er 2"* and *"low sulphur fuel"* respectively and on the other hand, it shows how a changeover to this latter fuel would increase the useful lifetime of some of those equipments directly involved such are heat exchangers, reducing substantially the cost of maintenance due to the considerable decreasing of the surface and time exposed to acid depositions with the subsequent increasing in boiler's performance and reliability.

The second part is focused on new studies on the effect that gradual replacement of *"fuel N^er 2"* by *"low sulphur fuel"* would have with regard to the environment. Thus, an analytical comparison (in terms of emissions) was carried out between both types of fuel-fired power plant schemes, showing this way the important advantages of the last ones, especially in terms of emissions and efficiency. Finally, a comparison with experimental values available from the gas analyzer has also been performed, showing a very good agreement with the data previously obtained in the calculations carried out here, so in this way that analytical procedure could be considered as validated. We will also evaluate the re-turning fuel oil boilers into natural gas ones which has risen over the last years. In this way, conventional fuel oil fired power plants are gradually replaced by natural gas-fired combined cycles, reducing sensitively the CO_2 emissions while SO_2 emissions are almost negligible. European Directives will be summarised showing the ceiling emissions.

[*] Corresponding author, Phone: + 34 946014250; Fax: + 34 946014283; E-mail: jesusmaria.blanco@ehu.es

Finally, the later study addresses the experimental optimization of a specific operation mode recently carried out over a thermal power plant working at minimum technical load (130 MW) by using a profitable combination of low-sulphur fuel burners and natural gas igniters, traditionally used as combustion support, analysing the economic and environmental advantages, considering also the fuels prices fluctuations in the context of the global market.

Keywords: Acid rain, SO_x emissions, Thermal power plant, efficiency, low-sulphur fuel, natural gas.

1. Introduction

There is a global concern about acid rain all over the world as we can experience some problems directly derived from it, such are severe deforestation, contamination of some water springs and so on. Acid rain is mainly caused by some combustion exhaust gases, especially NO_x and SO_2, but whereas NO_x can be reduced by as much as 75% through the use of, for example, low NO_x burners, staggered combustion technologies, [1-5], etc, SO_x depends mainly on the sulphur content in the matrix of the fossil fuel burned [6].

According to figure 1 the main process taking part in the acidification of earth are NO_x and SO_2 from human sources, such as electricity generation, factories and motor vehicles. Coal power plants are one of the most polluting. The gases can be carried hundreds of kilometers in the atmosphere before they are finally converted to acids and deposited. Factories used to have short funnels to let out smoke, but this caused many problems, so now, factories have longer smoke funnels. The problem with this is those pollutants get carried far off, where it creates more destruction.

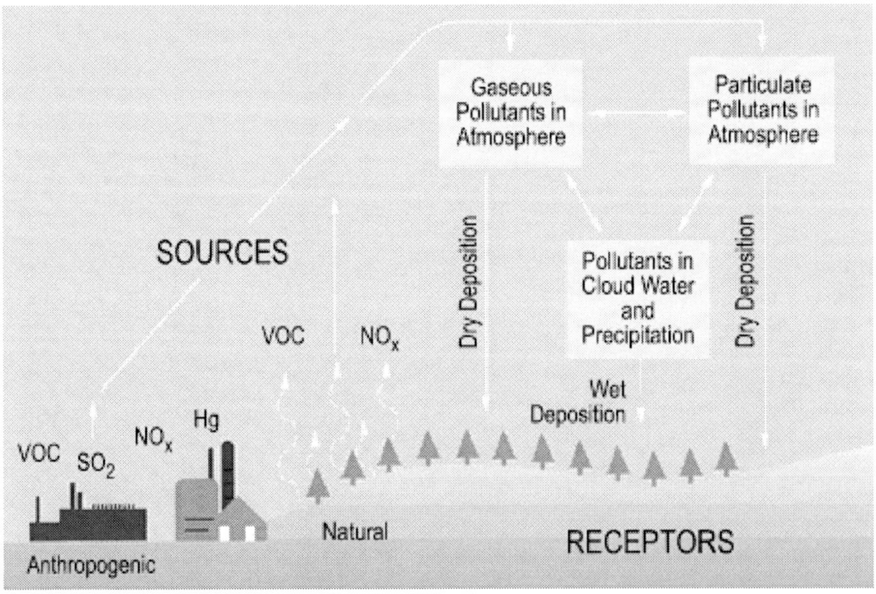

Figure 1. Main processes involved in acid deposition (courtesy of Wikipedia).

The term "acid rain" is sometimes used more generally to include all forms of acid deposition both wet deposition, (where acidic gases and particles are removed by rain or other precipitation), and dry deposition (removal of gases and particles to the Earth's surface in the absence of precipitation). Acid rain occurs when sulfur dioxide and nitrogen oxides are emitted into the atmosphere, undergo chemical transformations, and are absorbed by water droplets in clouds. The droplets then fall to earth as any kind of rain. This increases the acidity of the soil, and affects the chemical balance of rivers, water reservoirs and streams so it can even kill fish. Acid rain is a serious environmental problem accelerating weathering in carbonate rocks. It also contributes to acidification of forest at high elevations. High altitude forests are especially vulnerable as they are often surrounded by clouds and fog which are even more acidic than rain. Acid rain can slow the growth of forests, cause leaves and needles to turn brown and fall off and die. In extreme cases trees or whole areas of forest can die. The death of trees is not usually a direct result of acid rain; often it weakens trees and makes them more susceptible to other threats.

Some scientists have suggested direct links to human health, but none have been already proven. However, fine particles, a large fraction of which are formed from the same gases as acid rain (sulfur dioxide and nitrogen dioxide), have been shown to cause illness and premature deaths such as cancer and other deadly diseases.

Acid rain can also cause damage to certain building materials and historical monuments. Acid rain can cause weathering on ancient and valuable statues and has caused considerable damage. This is because the sulfuric acid in the rain chemically reacts with the calcium compounds in the stones (limestone, sandstone, marble and granite) to create gypsum, which then flakes off. This is also commonly seen on old gravestones where the acid rain can cause the inscription to become completely illegible. Acid rain also causes an increased rate of oxidation for iron [7-8]. Visibility is also reduced by sulfate and nitrate in the atmosphere.

Acid rain is measured using a scale called "pH." The lower a substance's pH, the more acidic it is. The pH scale measures how acidic or basic a substance is. It ranges from 0 to 14. A pH of 7 is neutral. A pH less than 7 is acidic, and a pH greater than 7 is basic. Each whole pH value below 7 is ten times more acidic than the next higher value. For example, a pH of 4 is ten times more acidic than a pH of 5 and 100 times (10 times 10) more acidic than a pH of 6. The same holds true for pH values above 7, each of which is ten times more alkaline (another way to say basic) than the next lower whole value. For example, a pH of 10 is ten times more alkaline than a pH of 9. A substance that is neither acidic nor basic is neutral.

Protection of the atmosphere is an environmental issue that will dominate Community policies in the coming century, affecting policies as diverse as energy, transport and land development. A number of international treaties on the long range transport of atmospheric pollutants have been agreed e.g. Sulphur Emissions Reduction Protocol under the Convention on Long-Range Tran boundary Air Pollution. A more recent regulatory scheme such the Kyoto protocol, involves emissions trading. In this scheme, every current polluting facility is given an emissions license that becomes part of capital equipment. Operators can then install pollution control equipment, and sell parts of their emissions licenses. The intention of this is to give operators economic incentives to install pollution controls.

The overall goal of the US EPA (Environmental Protection Agency) program is to achieve significant environmental and public health benefits through reductions in emissions of sulfur dioxide (SO_2) and nitrogen oxides (NO_x), the primary causes of acid rain. To achieve this goal at the lowest cost to the public, the program employs both traditional and innovative, market-based approaches for controlling air pollution. Specifically, the program seeks to limit, SO_2 emissions from power plants at 8.95 million tons annually starting in

2010, authorizes those plants to trade SO_2 allowances, and reduces NO_x emission rates. In addition, the program encourages energy efficiency and pollution prevention [9].

The European Parliament has played a decisive role in the formulation of a progressive environmental policy to combat air pollution. After successfully forcing through stricter limit values for motor vehicle exhaust fumes despite the opposition of the Commission and Council, it called in another resolution for a further reduction in the limit values for pollutant and noise emissions from cars and aircraft and for stricter fuel quality standards.

For instance fuel oils, widely used in thermal power plants, contain a certain amount of sulphur in variable percentage depending on the type of fuel oil [10-11]. During the combustion process, the sulphur present in the fuel will be rapidly combined with oxygen to form sulphur dioxide (SO_2) in a partial oxidation and sulphur trioxide (SO_3) in a full oxidation respectively [12]. It is the presence of these generic sulphur oxides (SO_x) in the flue gas that represents the largest potential cause for acid rain [13], but also for extra corrosion of the equipment involved when the acid dew point temperature (ADT) is reached.

Thus, SO_2 could be dissolved in any free moisture that may be present in the flue gas to form sulphurous acid (H_2SO_3) a powerful corrosive agent. Approximately 1 to 2% of the SO_2 will be further oxidized into SO_3, which combined with any water flue, forms sulphuric acid (H_2SO_4). Consequently, this can only occur at temperatures lower than the water dew point (referred to in some literature as "physical dew point temperature"), [14] which is the result of partial steam pressure on combustion gases and is not therefore likely to occur at temperatures higher than 60 °C. In other words, there is a possibility of dissolving SO_2 in water, if use is made of the steam's latent heat of condensation contained in combustion gases. These condensations take place with concentrations of around 98% and in these circumstances are not very aggressive. However when boilers are shut down, the highly water-repellent H_2SO_4 adsorbs atmospheric water vapour and reaches concentrations of around 40 %, whereupon it becomes especially aggressive with regard to any metal surface.

The first part of this work summarizes a method for determining the ADT in a reference thermal power plant for two types of fuel, such as: "*fuel N^er^ 2*" and "*low sulphur fuel*" respectively. The first one, widely used in Spanish thermal power plants, has a mass percentage of sulphur of around 2.9 % whereas the second one should have < 1% (in fact, it tends to be around 0.7%). This part also shows that replacement of "*fuel N^er^ 2*" by "*low sulphur fuel*" will significantly reduce maintenance on the equipments, due to the considerable reduction in the time exposure to further extra corrosion or acid corrosion, so there will even be a possibility of withdrawing a number of other air pre-heaters from service which, in view of the extra consumption of steam in the boiler, are used to increase the "cold end" temperature of the regenerator, [15-16] with the subsequent improvement in boiler performance and reliability. All acids are corrosive to steel, i.e. tubing, casing, tools, etc. As the temperature in the well bore increases the corrosion rate increases, sometimes dramatically, depending on the acid strength and system used in the application. For this reason, acid corrosion inhibitors are mandatory for all acidizing treatments.

The second part is focused on the effect that gradual replacement of "*fuel N^er^ 2*" by "*low sulphur fuel*" would have now with regard to the environmental impact. Thus, an improved analytical comparison (in terms of emissions) was carried out between both types of schemes showing numerically the great advantages of the last ones.

A comparison with experimental values available has also been performed [17], showing a very good agreement with the data obtained with the calculations carried out here, so this analytical procedure can be considered validated. Flow measurement to determine the exact quantity of pollutants produced is becoming more common. Many instruments are being equipped for this purpose now. Flow measurement allows the total level of pollution to be calculated and also controlled.

From the decade of the 90´s, the possibility of re-turning fuel oil boilers into natural gas ones raised, due mainly to the increasing of environmental concern about the "greenhouse effect" [18-20]. Nowadays, conventional power plants are gradually replaced for natural gas-fired combined cycles, due to important advantages with regard to the environmental impact, reducing sensitively the CO_2 emissions while SO_2 emissions are almost annulled [21-23]. We will also evaluate this changeover in terms of emissions. The use of natural gas is proposed because its basic component (CH_4), only contributes to the greenhouse effect in about 18 % and SO_2 emissions are almost negligible, just due to the addition of some odorants such as "t-butyl mercaptan", with a rotting-cabbage-like smell, is added to the otherwise colorless and odorless gas, so that leaks can be detected before a fire or explosion occurs.

Finally, the last study is focused on an experimental way to reduce acid emissions in the power plant of reference working at minimum technical load (130 MW). Traditionally, this involved operating the boiler by firing the 7 pairs of low-sulphur fuel oil burners as well as the 2 pairs of natural gas igniters in order to ensure the burners start-up, shut-down and venting procedures respectively [24]. We shall henceforth refer to this conventional mode of operation as "alternative A".

Thus, the procedure proposed in this study in order to minimize acid emissions consist of testing different combinations of fuel oil and natural gas burners working together, so as a result of this, the best option selected from amongst all those tested was to operate the boiler by firing up 6 pairs of low-sulphur fuel oil burners, plus the whole of the 12 pairs of natural gas igniters available, which besides ensuring the operating safety and requirements of the low sulphur fuel burners would operate as real natural gas burners. We shall refer to this mode of operation as "alternative B".

This study addresses the optimisation of one particular operation mode of a thermal power plant using this specific combination of low-sulphur fuel oil burners and natural gas igniters, analysing this way the economic and environmental advantages of the proposed solution.

2. Aims and Methodology

The aim of the present work is to optimize the plan operation scheme in any fossil-fired power plant in order to minimize the SO_x emissions that are directly responsible for the "acid rain effect", showing this way the latest efforts made in order to evaluate acid dew point temperature, diminishing acid emissions, as a result of both changing fuels and power schemes, and finally showing experimental results from test carried out over a specific thermal power plant.

In this way the methodology can be summarised as:

- Improvements in the determination of the ADT, including the influence in the cycle life of a regenerator, directly exposed to acid depositions for different fuels.

- Analytical evaluation of the acid emissions over the thermal power plants depending on the fuel used and comparative test between conventional and combined natural gas powered cycles.
- Implementation of an experimental methodology for optimizing the working conditions when the plant is operating at minimum technical load especially in terms of emissions and cycle efficiency. Further economical considerations will be taking into account with regard of the fluctuations of fuel prices.

3. Description of the Experimental Facilities

Both regenerator and gas analyzer will be described next, addressing a brief description of their principles and other important considerations.

3.1. The Rotary Regenerative Heat Exchanger

Figure 2 shows the layout of the power plant of reference, corresponding to a regenerator (rotary regenerative type) where the heat transfer process occurs between acid hot gases coming from the economizer and fresh air in order to send this last one into the boilers as a preheated combustion air. Places where both experimental temperatures and pressures probes were situated are also shown in this figure.

The basic component of this type of rotary regenerative heat exchanger is a continuously rotating cylinder, called rotor, which is packed with thousands of square thin-walled steel plates, packed in frame baskets of heat transfer surface. When the exhaust gases leave each one of the air heaters in this installation, do not go directly to the stack, but first pass through the respective mechanical precipitators where the ash and cinders in the gases are retained by means of centrifugal forces created by settling cyclones, and collected later by a deposition tank through a deposition valve.

During each revolution of the rotor, heat is absorbed by the heating surface passing through the hot gas stream and transferred to the combustion air stream. This accumulated heat is released into the incoming air as the same surfaces pass through the other half of the structure. The heat transfer cycle is continuous in periods of about 60 seconds as the surfaces are alternately exposed to the outgoing gas and incoming air streams [25-26]. Heat energy is captured and transferred to the incoming air for combustion before it is lost to the stack. The result is a substantial saving in fuel that would otherwise be required to bring the air up to combustion temperature [27]. Figure 3 shows principles of the "Ljungström" Air Preheater.

Until now, such regenerators were designed to rotate steadily and continuously, being used mainly for gases near atmospheric pressure. They are used in boilers as air preheaters, in flue gas desulphurization plants as gas/gas heaters and in tail-end selective catalytic reduction NOx plants as gas pre-heaters. They are both designed and built to operate over extended periods with durable and uninterrupted service. Simplicity of design also makes them easy and economical to maintain while in operation and at scheduled outages [28-30]. The majority of air pre-heaters in use in the power industry today are of the "Ljungström" type, which is a regenerative heat exchanger, invented in the 1920s and soon used throughout the world. Dr. Fredrik Ljungström, then technical director at "Aktiebolaget Ljungström

Angturbin", invented it for preheating combustion air in boiler plants, but the use has expanded to include energy recovery in combination with the removal of oxides of sulphur and nitrogen.

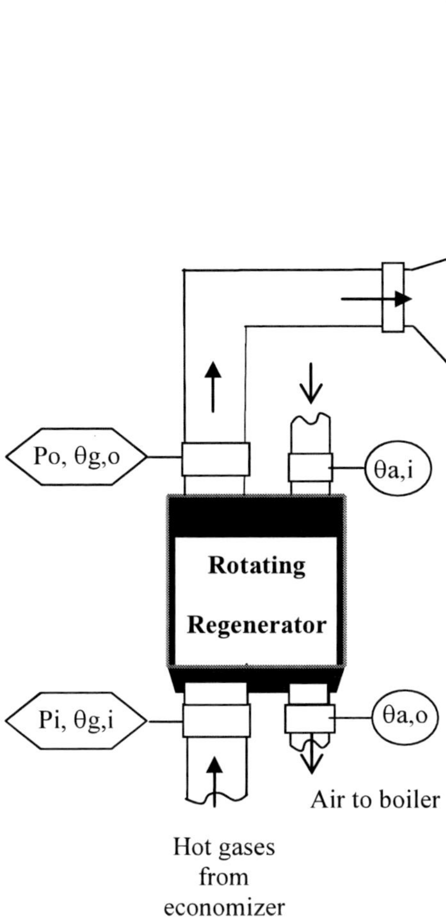

Figure 2. Layout of the power plant corresponding to the regenerator.

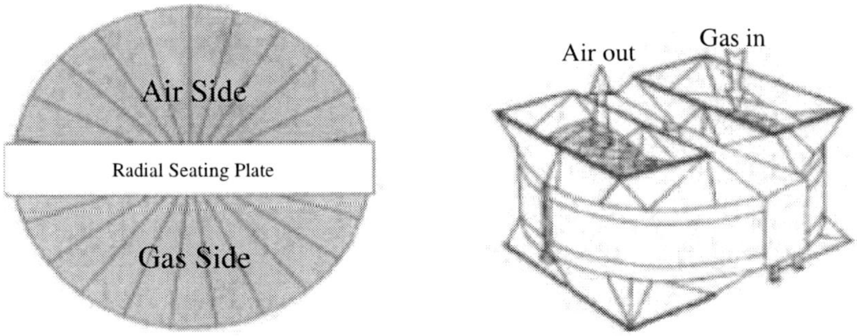

Figure 3. Principles of the Ljungström Air Preheater.

Throughout the history of boilers, much advancement have obtained better performance and lower fuel consumption, but few have been as remarkable or have been adopted so pervasively as the "Ljungström" air preheater. For example, just the first installation saved as much as 25 percent of the fuel consumption. In a modern utility boiler, it provides up to 20 percent of the total heat transfer in the boiler process, while only representing 2 percent of the total investment. An estimate of total worldwide fuel savings is equivalent to 4,960 million tons of oil. In the United States (US), it annually saves about $30 billion US dollars.

The earliest installations have been taken out of service because the boiler plants were torn down. Thus the landmark is an early working model used in experiments from 1961 to 1977 at "Svenska Rotor Maskiner AB", now featured in an exhibit at the "Tekniska Museet". Several inventions led to development of the air preheater, beginning in 1908 when brothers Birger and Fredrik Ljungström founded "ALA" to produce double rotating steam turbines. The first tests of the air preheater were performed in ALA's facilities in Gashaga, Sweden. The prototype was later installed at the factory boiler there. The first commercial installation was made in 1922 for "AB Forenade Choklad Fabrikerna" (a famous chocolate manufacturer) on the island of Kungsholmen in Stockholm.

In 1995, the American Society of Mechanical Engineers (ASME) declared them to be a "Mechanical Engineering Landmark" [31].

3.2. The Gas Analyzer

In direct measurements, the temperature values and also the concentration of those gas elements which are detected by independent electrochemical sensors can be obtained. The electrochemical cell indications are proportional to the volume concentration of the detected elements, expressed in ppm (parts per million).

The following quantities are obtained by means of direct measurement:

- Flue gas temperature (°C)
- Ambient temperature (°C)
- Volume concentration of CO (ppm)
- Volume concentration of NO (ppm)
- Volume concentration of SO_2 (ppm)
- Volume concentration of O_2 (%)

Ambient temperature is mostly measured using a thermistor. This is a material that changes its resistance proportional to the temperature. The commonest type is probably the platinum thermistor. Flue gas temperature is generally measured with a thermocouple. A thermocouple produces a tiny electrical potential proportional to a temperature difference between two points. There are different types of thermocouple available for different temperature ranges, but most of them cover the combustion gas range.

With these main factors all the other parameters for flue gas are calculated. There are certain sensors that have a different method of measuring to the standard volume concentration indication generally used. This is particularly common with oxygen. Here, the sensor shows the partial pressure of the gas instead of a volume concentration.

Naturally it is then necessary to know the ambient pressure in order to calculate the volume concentration if needed. Generally it is enough to know the height above sea level and to calculate the ambient pressure on this basis. These sensors are used in medical equipment where the partial pressure reading is most important, hence the difference in technology. More accurate readings will require exact knowledge of the ambient pressure, however. This is also important when using infrared or similar technology. Here, the sensor basically "counts" the number of molecules in the light path, which will also be a function of pressure. More sophisticated infrared analysers are therefore also fitted with a sensor for ambient pressure to correct for these effects.

The range from a simple panel containing a knock-out pot and silica gel drier, (only suitable for very short term operation without the risk of damage to the analyser and not at all suitable for the measurement of soluble components such as NO_2, SO_2 and HCl) to mains powered units which combine “Peltier” coolers, filters, sample and condensate pumps and simple sample probes. More sophisticated units are becoming available which include heated sample probes and heated sample lines. The permeation dryer is also becoming interesting, since measurements with infrared sensors require that the water be completely removed from the system. Even low levels of water vapour will produce false peaks on the sulphur dioxide and nitric oxide measurements.

Most of these sample conditioners, with good portable flue gas analysers, are very useful tools for checking permanent installations, for use where permanent analysers are not justifiable, or technical problems exist in a particular process or furnace prohibiting the installation of a permanent system. The success of sample systems and their associated analysers is entirely dependent on good maintenance. Any analyser system requires a maintenance schedule which must be adhered to. Experience shows that over 97 % reliability is achievable with good preventive maintenance. The fuel saved because of that can alone often pay for the analyser and its maintenance, not to mention the positive effects on our environment. The flue gas analyser is no longer so expensive that only huge companies can afford to buy and operate a combustion gas tester. When you add in the possibility of extra profit by so-called CO_2 trading as envisaged in the Kyoto Protocol, the reasons sum up in favour of testing. It is not a luxury, it is an essential.

Flue gas analysers have also been reduced dramatically in size from the old type of cased instrument weighing around 20 kg to an instrument that can truly be designated a hand held analyser. Weighing less than one kilogram in most cases they are powerful and less expensive relatively than they ever were. The flue gas analyser has really ceased to be an expensive laboratory instrument and is now a medium priced field instrument, in use on building sites and on the factory floor. Laboratory style flue gas analysers do still exist, of course, but they are more used for experimental work or as standard for comparison to other flue gas analyzers.

Overall, electronic flue gas analysers have many important advantages, such as ease of use and speed, automatic sampling, calculations and report generation. The ability to connect them to a computer and download stored results has made their use much quicker and simpler. Most common flue gas analysers have enough memory capacity to store the results from a day's work, which can then be collected in one package afterwards, complete with comments and date/time etc. Cordless communication is here also coming into use. A Bluetooth connection can easily be applied to an existing interface enabling wireless communication with a range of different instruments.

To perform an analysis, the equipment must be clean and operating properly. After you have serviced the equipment, let it run for 15 to 20 minutes. This will allow the unit to warm up and vent properly. This is especially important after cleaning soot out of the flue, since the particle left will burn inefficiently and will falsify the result. Each furnace manufacturer dictates how to do the actual testing on their products. If the service manual does not indicate how, refer to general guidelines set up by the manufacturer of the testing equipment you use. Most modern equipment will come with suggestions as to the pollutant limits typical for the system, so this should increasingly be less of a problem. In any case, the EPA (Environmental Protection Agency) or equivalent standard compliance limits represent a maximum level of pollution allowable.

In all cases, the flue temperature will be measured at the same point as the gas sample is taken. If there are sufficient heat exchangers obstructing the stack, then the gas may be almost homogenous, but care should be taken to check the gas stream for the presence of a core of high temperature gas with a differing composition. This is the gas which should be sampled, both for pollutants and for the true flue gas temperature. The sample point should be before draft diverters and barometric dampers so that the gases are not diluted and the temperature has not been decreased by the addition of outside air at least 15 cm upstream from the furnace side of the draft regulator (diverter), and as close to the furnace breeching as possible.

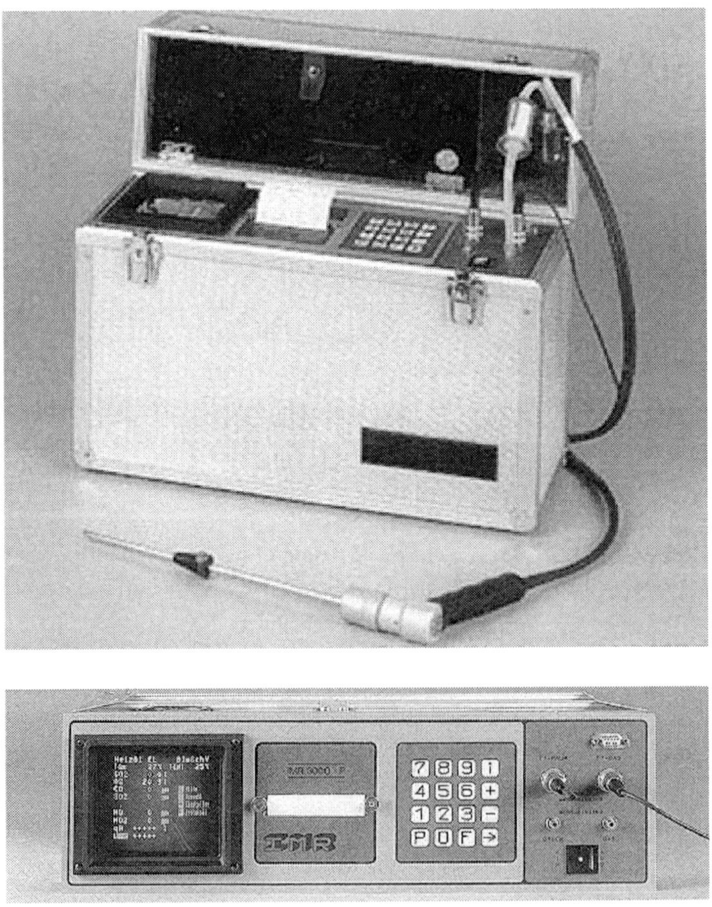

Figure 4. Portable IMR3000 gas analyzer with the aluminium enclosure and detail of the front panel.

Although some combustion experts do weekly tests others consider three or four checks a year an acceptable level of testing. The appropriate level for a particular unit can only be determined by experience, and is partly dependent on the fuel consumption of the burner. The higher the consumption, the more important it is to optimise the process. Fuel consumption will be affected by a number of factors that can be measured with a flue gas analyser. The air/fuel ratio is an obvious candidate, together with the pressure drop in the stack. The carbon monoxide produced and a few other factors all add up together to change the efficiency of the burner. What works at low load my not work so well at full or half load.

Our portable gas analyzer, depicted in figure 4, is a microprocessor-based gas analyzer. It has a totally automatic microprocessor-based operation including auto start-up, auto-calibration, auto-range, internal parameter monitoring and alarms.

This makes the instrument ideal as a "smart" stand-alone unit or as a "building block" for systems integration via communications RS 232 C port or remote contact closures. Automatic diagnostics and fully-configurable concentration alarms are also integral to this gas analyzer. A large multifunction 5" LCD monitor provides a clear display of carbon monoxide, carbon dioxide, NO or sulphur dioxide gas concentration measurement, range, sample flow rate and the status of critical operating parameters including sample flow, pressure, temperature and DC power supplies. The intuitive, menu-driven interface and soft key buttons enable the operator to quickly configure and control the gas analyzer function. The unit features a paramagnetic oxygen measurement within the analyzer. The main technical specifications for this analyzer can be seen in table 1.

Table 1. Main specifications for the portable IMR3000 gas analyzer.

Response Time	5 s
Accuracy	+/- 1.0 %
Linearity	0.5 %
Max. Temp.	1,500 °C
Voltage	230 V / 50Hz, 110 V / 60Hz
Battery	12 VDC (6h autonomy)
Interface	RS 232 C
Memory	10,485 measurements
Monitor	5" LCD
Units	ppm, mg/Nm^3
Emissions	CO, CO_2, SO_2, O_2, NO, NO_2, H_2S
Dimensions	510 x 180 x 430 mm
Weight	Aprox. 16 kg

4. New Features in Determination of the Acid Dew Point Temperature

4.1. Definition and General Considerations about ADT

When the current acid flow is constant, the rate of condensation is equal to the rate of evaporation. The temperature at which this occurs is the acid dew point temperature (ADT), which is a direct measurement, requiring no calibration or reference. Whenever tube wall surfaces in boilers air heaters or economizers fall below acid dew point temperatures of vapours such as hydrochloric acid, nitric acid, sulphuric acid, condensation of these vapours can occur on these surfaces, leading to extra corrosion and failures which can even collapse the equipment directly exposed.

There have been many developments in the field of the combustion technology in the last few years. This is especially true for gas burners. The newest gas burners are of the condensing type, where the flue gas temperature is below the dew-point temperature of the gases, which lies by about 57 °C for flue gases from gas burners. The water in the gas condenses on a set of special heat exchangers mounted in the gas stream. This enables the evaporation energy to be extracted from the gas as well, raising the efficiency greatly at no extra cost. Care has to be taken to use corrosion-resistant materials, since the condensate is relatively aggressive and the flue channel has to be fitted with a stainless steel liner or equivalent, otherwise it will be destroyed fairly quickly.

The traditional calculation of combustion efficiency uses the temperature of the gases at inlet and in the stack. A separate correction must be made to take account of the specific heat of evaporation when working with condensing equipment. This correction is of interest for the owner of the equipment, but may not be accepted officially as a measure of efficiency. Regulations in most countries specify how these parameters should be calculated and may not allow a different procedure.

In the combustion of standard fuel-oils in Spanish power plants, the concentration of SO_2 in terms of mass and volume in combustion products tends to be between 1,400 and 7,000 mg/Nm^3 (500 ppm and 2,500 ppm. respectively). Figure 5 has been obtained experimentally over the power plant of reference. It shows the formation of the ADT over clean metal plates vs. the SO_2 concentration (mg/Nm^3) in the combustion exhaust gases for a range of the above mentioned values. According to this curve, it can be deduced that ADT values are commonly in the range of 125 and 145 °C, depending on the SO_2 amount.

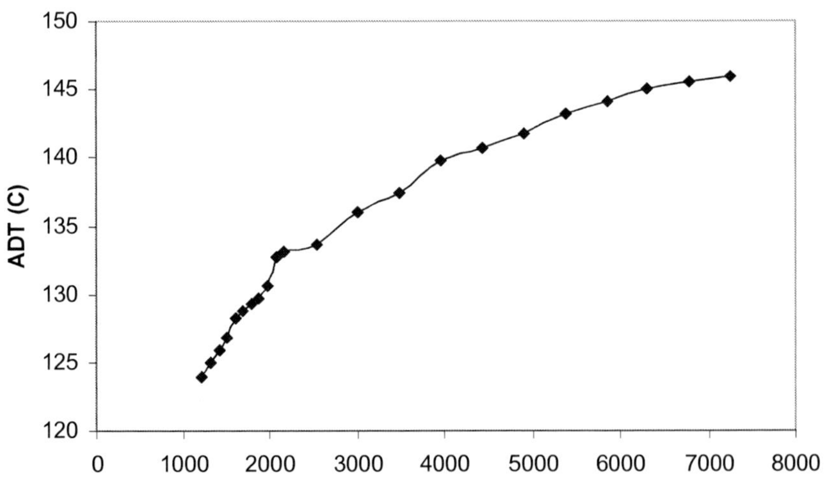

Figure 5. Experimental curve of ADT (°C) vs. $[SO_2]_m$ (mg/Nm^3).

There follows a discussion of the specific results for each one of the two fuels considered. The salient data for "*fuel N^er^ 2*" (LCV = 40,035.66 kJ/kg fuel) and "*low sulphur fuel*" (LCV = 40,630.68 kJ/kg fuel) are featured in Table 2 [23].

Table 2. Mass composition of a "*fuel N^er^ 2*" and "*low sulphur fuel*" and their main combustion reactions.

% mass	fuel N^er^ 2	low sulphur fuel	Basic combustion reactions (*)
C	85.40	85.98	$C + O_2 \rightarrow CO_2$
H	10.38	10.91	$H_2 + \frac{1}{2}O_2 \rightarrow H_2O$
S	2.92	0.81	$S + O_2 \rightarrow SO_2$
N	0.55	0.46	-
O	0.00	1.81	-
H_2O	0.45	0.00	-

(*) The amount of CO was negligible, according to the data provided by the gas analyser.

4.2. ADT Burning "Fuel N^er^ 2"

Via combustion calculations for our reference power plant, we have obtained both the mass and volume concentration of SO_2 in the combustion gases when burning "*fuel N^er^ 2*". Nevertheless, due that the gas analysers do not provide mass results (they receive gases already cooled, so the analysis is performed on dry gases), this will be the basic value used for ensuring consistency with the data from the combustion gas detectors [32-33]. Thus:

$$m_{SO2\,(DG)} = 0.0583 \text{ kg/kg fuel} \tag{1}$$

$$V_{SO2\,(DG)} = 10.7928 \text{ Nm}^3\text{/kg fuel} \tag{2}$$

$$[SO_2]_m = \frac{0.0583}{10.7928} \cdot 10^6 = 5{,}402 \text{ mg/Nm}^3 \tag{3}$$

Considering the value obtained in eq. (3) in figure 5, the ADT finally obtained for "*fuel N^er^ 2*" is:

$$\underline{\underline{\theta_{ADT,FN2} = 143 \text{ °C}}} \tag{4}$$

It should be noted how SO_x emission values may be rigorously obtained by means of combustion calculations [18]. Anyway in this case, this calculation value has been compared with that provided by the combustion gas analyser in the power plant of reference and the result is very similar. This has enabled us to verify the analyser (correct sitting of the probe, etc.).

4.3. ADT Burning "Low Sulphur Fuel"

When burning *"low sulphur fuel"*, there are no analytical data on the mass concentration of SO_2 in dry combustion gases, because the relevant combustion calculations have not been carried out. However, we do have the value provided by the combustion gas analyser, which is acceptable, as its measurement results have been considered validated, as mentioned above.

$$[SO_2]_m = 1{,}300 \text{ mg/Nm}^3 \quad (5)$$

Considering the value obtained in eq. (5) to figure 5, the ADT finally obtained for *"low sulphur fuel"* is:

$$\underline{\underline{\theta_{ADT,LS} = 124\ ^\circ C}} \quad (6)$$

It simply means that burning *"low sulphur fuel"* the possibility of reaching ADT is lower.

5. Evaluation of the Equipment Directly Exposed to Acid Depositions

In recent years, with the drive for greater profitability, the process industries have been placing greater emphasis on improving process performance, operating closer to constraints and reducing downtime. Several technologies, such as process modelling, statistical process control, optimisation techniques and advisory systems have helped to make these goals a reality [34].

Difficulties are encountered when process industries begin to deal with optimisation tasks such as minimising total cost and maximising availability [35]. In order to determine optimal maintenance, the effect on plant performance of alternative maintenance policies and the costs of the resources they would require must be compared with alternative uses of these same resources. This requires a full scale detailed analysis if the best policy is to be selected [36].

5.1. Considerations on Time Dependence Calculations

The heat transfer process between the air and gas zones in the rotary regenerators (Ljungström type), responds to the case of walls subject to regular changes in surface temperatures [37-38]. Thus, the time dependence temperature all over the metal panels can be deduced from equation (7), corresponding to one-dimensional conduction in transient state.

$$\frac{\partial \theta}{\partial t} = a \cdot \frac{\partial^2 \theta}{\partial x^2} \quad (7)$$

The solution of this differential equation provides the following expression (8) for the periodic time dependence temperature on a plane within the wall at a distance of "x" considering oscillation around an average temperature (θ_{av}) [39].

$$(\theta - \theta_{av}) = (\theta_{máx} - \theta_{av}) \cdot e^{-x\sqrt{\frac{\pi}{aT}}} \cdot \cos\left[x \cdot \sqrt{\frac{\pi}{a \cdot T}} - \frac{2\pi}{T} \cdot t\right] \quad (8)$$

And the expression for the surface temperature distribution (x = 0) adopt a sine-wave function shape:

$$(\theta - \theta_{av}) = (\theta_{máx} - \theta_{av}) \cdot \cos\left[-\frac{2\pi}{T} \cdot t\right] \quad (9)$$

In practice, the trend in the surface temperature of the metallic panels that make up the baskets (heat transfer panels) on the air heaters does not looks like the one we have just addressed, but rather a parabolic-type function in heating and similar to a logarithmic-type function in cooling, which has been addressed experimentally, as may be seen in figure 6, addressing both time and sector of baskets in "x" axis.

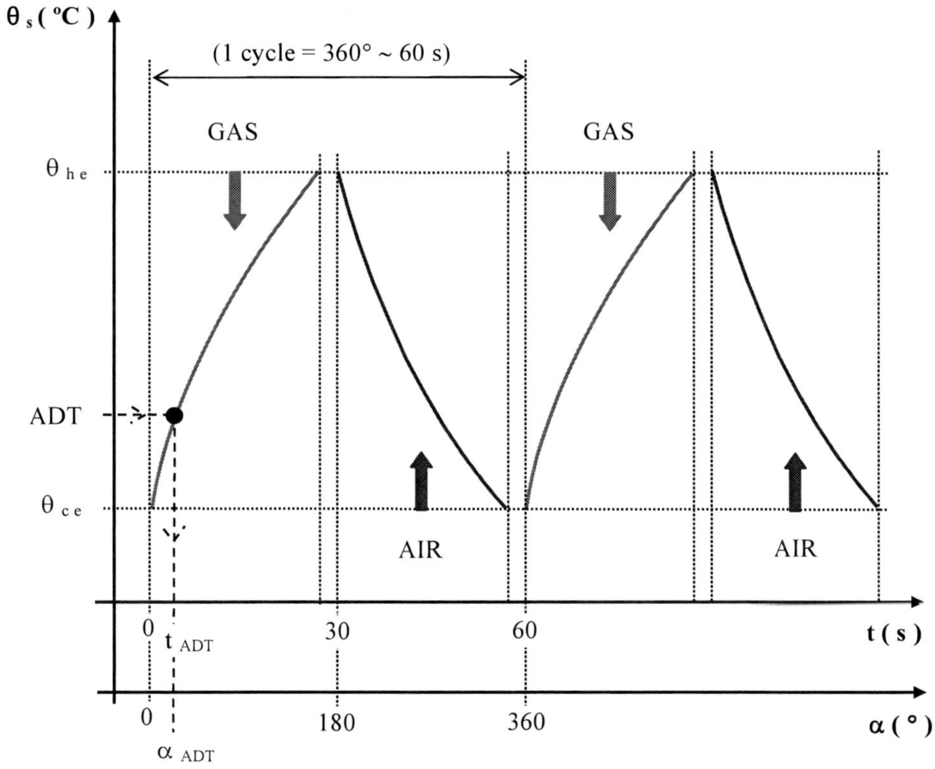

Figure 6. Empirical time-angle dependent surface temperature over a rotary regenerative heat exchanger.

The temperature of the panels on the heater baskets obviously cannot be measured directly either with thermocouples or with other direct contact devices, as the panels have a uniform circular movement. Nevertheless, it can be ascertained with sufficient accuracy by means of the algebraic sum of the air and gas temperature readings [40-41] on each side of the mechanical seals on the rotating regenerators, in the air outlet and combustion gases inlet zone that constitutes the lowest temperature area over the panels, referred to as the "cold end temperature" ($\theta_{c,e}$) [42], and on the combustion gases outlet and air inlet zone that records the highest temperature on the basket panels, known as the "hot end temperature" ($\theta_{h,e}$).

In order to prevent cold end corrosion and air heater fouling it is important to ensure that the flue gas temperature is above the sulphuric acid dew point temperature. However, if the temperature is too high then energy is wasted and plant efficiency is reduced. In order to operate at the optimum point, a continuous sulphuric acid dew point monitor should be employed for this matter. The output from the monitor can be used within the control loop, or by manual trim, for maintaining the optimum flue gas temperature. Direct measurement, giving accurate and reliable monitoring of the sulphuric acid dew point temperature will assist with on-line control of flue gas temperatures, minimising maintenance costs and improving the total efficiency of the process.

According to figure 6, the time per cycle that metallic plates are directly exposed to acidic depositions can be calculated very easily as well as sectors of baskets directly exposed to extra acidic corrosion. There follows again the specific procedure for each one of the two fuels.

5.2. Burning "Fuel N^{er} 2"

The average measurements of a high number of temperature readings for the power plant of reference operating at full load with "*fuel N^{er} 2*" provide the values which are used to obtain both temperatures $\theta_{c,e}$ and $\theta_{h,e}$ respectively as it is shown in table 3.

Table 3. Cold end and hot end temperatures for "*fuel N^{er} 2*".

fuel N^{er} 2	
Cold end values	**Hot end values**
$\theta_{a,i} = 43.0$ °C	$\theta_{a,i} = 286.5$ °C
$\theta_{g,o} = 156.5$ °C	$\theta_{g,o} = 364.5$ °C
$\theta_{c,e} = \dfrac{\theta_{a,i} + \theta_{g,o}}{2} = 99.75\ °C$	$\theta_{h,e} = \dfrac{\theta_{a,i} + \theta_{g,o}}{2} = 325.5\ °C$

According to values obtained from this table, the time directly exposed to acidic corrosion of the metallic plates for each period of 60 s. when burning *"fuel N^{er} 2"* is:

$$\underline{\underline{t_{ADT,FO2} = 5.74\ s}} \tag{10}$$

Consequently, keeping the pre-heaters in service, the sector (α_1) of baskets exposed to ADT works out finally as follows:

$$\underline{\underline{\alpha_1 = 34.42\ ^\circ}} \tag{11}$$

5.3. Burning "Low Sulphur Fuel"

The average measurements operating at full load with "*low sulphur fuel*" provide the values which are used to obtain both temperatures $\theta_{c,e}$ and $\theta_{h,e}$ respectively as it is shown in table 4.

Table 4. Cold end and hot end temperatures for "*low sulphur fuel*".

low sulphur fuel	
Cold end values	**Hot end values**
$\theta_{a,i} = 32.5$ °C	$\theta_{a,i} = 287.0$ °C
$\theta_{g,o} = 149.5$ °C	$\theta_{g,o} = 365.0$ °C
$\theta_{c,e} = \dfrac{\theta_{a,i} + \theta_{g,o}}{2} = 91.0\ ^\circ C$	$\theta_{h,e} = \dfrac{\theta_{a,i} + \theta_{g,o}}{2} = 326.0\ ^\circ C$

According to values obtained from last table, the time directly exposed to acidic corrosion of the metallic plates, for each period of 60 s. when burning *"low sulphur fuel"* is:

$$\underline{\underline{t_{ADT,LS} = 4.21\ s}} \tag{12}$$

Consequently, keeping the pre-heaters in service, the sector (α_2) of baskets exposed to ADT works out finally as follows:

$$\underline{\underline{\alpha_2 = 25.56\ ^\circ}} \tag{13}$$

So it can be deduced that time directly exposed to acidic corrosion of the metal plates, (for each period of 60 s.), is lightly lower when burning *"low sulphur fuel"*. Besides, as the amount of these acid emissions is lower when using last mentioned fuel, so a considerable reduction in the sectors of the baskets directly affected by the acid depositions can be expected.

This is a very important goal, with relation to the maintenance cost, bearing in mind that the manufacturer of the rotating regenerators recommends checking and replacing baskets every 4 years when burning "*fuel N^{er} 2*". According to the experience the fact of replacing this fuel with "*low sulphur fuel*" would extend the maintenance time by around 1 year more, thereby leading to a considerable extra financial saving.

5.4. Protection against Acid Corrosion

Corrosion of iron and steel substrates can be inhibited by coating them with any of the wide variety of compounds denoted generally as polyanilines. A polyaniline suitable for this type of application can be in either an electrically conductive salt (doped) form or an electrically nonconductive base form. Typically, polyaniline is dissolved in an organic solvent and the resulting solution is applied to a substrate by spraying, dipping, or brushing. The solvent is then allowed to evaporate leaving the substrate coated with a solid film of polyaniline, typically 1 to 200 μm thick [43].

In general, the synthesis of a polyaniline involves mixing an aniline monomer with a protonic acid (e.g., hydrochloric acid), and a polymerization agent in an aqueous medium. The mixture is stirred and maintained at a temperature that depends upon the specific formulation and is typically a few degrees above or below 0 'C. The polymer forms as a blue-green precipitate, which is collected.

The polyaniline precipitate is washed with a protonic acid and/or organic solvent to obtain a polyaniline salt with a molecular structure like that shown in the second part of the figure. If desired, the polyaniline salt can be treated with an aqueous base (e.g., ammonium by droxide) to convert the salt to a base with a molecular structure like that shown in the third part of the figure. The resulting polyaniline base is preferably in an oxidation state associated with a form of polyaniline called "emeraldine". The polyaniline base can then be recovered by such established techniques as washing with an aqueous base and organic solvents, followed by vacuum drying [44-45].

Preferably, a coating solution is made from a polyaniline in the base form because of the relatively high solubility of the base form. Solvents that can be used to make polyaniline base solutions include dimethyl sulfoxide, N-methylpyrrolidinone, and tetramethylurea. The concentrations of polyanilines in useful coating solutions lie approximately in the range between 0.5 and 5 weight percent.

Upon exposure to a corrosive, acidic environment, a polyaniline coating film in base form undergoes a conversion to a doped, electrically conductive salt form. The polyaniline film coating can also be applied to the substrate in a partially doped form; however, the partially doped polyaniline is generally more difficult to dissolve in a given solvent. Useful dopant anionic species that can be used to make partially doped polyaniline coatings include sulfonic acids and camphor sulfonic acid; solvents that can dissolve the resulting partially doped polyanilines include toluene and substituted phenols [46].

The tenacity of a polyaniline coating can be increased by pretreating the surface of the substrate with a suitable compound. Compounds useful for this purpose are some that simultaneously bind chemically, via covalent or ionic bonds, to the surface of the substrate and to the polyaniline. Among these compounds are phosphoric acid, polyphosphoric acid, and organic chelating agents in which atoms form multiple coordinate bonds with the substrate.

6. Evaluation of the Emissions over the Thermal Power Plants

6.1. Analytical Calculation

The rapid oxidation of the sulphur is not realized in a typical way, since in a complete combustion more than 90 % of the sulphur mass content does not oxidize totally to form SO_3 but the above mentioned oxidation should normally be partial or incomplete, producing SO_2.

The quantity of SO_3 formed, depends on several factors for the combustion process [47-48], such are:

- Excess of oxygen in the combustion.
- Temperature of the flame and,
- Presence of catalysts.

Even considering a good performance of the great capacity boilers, the amount of SO_3 formed, is just around 1.5 %, being its partial pressure which determines the formation of the ADT. Nevertheless, as far as SOx evaluation concern, it can be considered to be formed entirely for SO_2 [49].

Thus, for both fuels according to table 1, the number of moles of sulphur dioxide can be obtained from the combustion reactions as:

$$\left[\psi_{SO_2}\right]_{FO2} = \psi_S = 0.9110\,\mathrm{mol\,SO_2/kg\ FO2} \tag{14}$$

$$\left[\psi_{SO_2}\right]_{LS} = \psi_S = 0.2531\,\mathrm{mol\,SO_2/kg\ LS} \tag{15}$$

So the volume of SO_2 can be seen respectively in eqs. (16) and (17).

$$\left[V_{SO_2}\right]_{FO2} = \psi_{SO_2} \times \frac{22.412}{1{,}000} = 0.0204\ \mathrm{Nm^3SO_2/kg\ FO2} \tag{16}$$

$$\left[V_{SO_2}\right]_{LS} = 0.0056\ \mathrm{Nm^3SO_2/kg\ LS} \tag{17}$$

And therefore the mass of SO_2 (considering $M_{SO_2} = 64.06\ \mathrm{g/mol}$), can be seen respectively in eqs. (18) and (19).

$$\left[m_{SO_2}\right]_{FO2} = \psi_{SO_2} \times M_{SO_2} = 0.0583\ \mathrm{kg\ SO_2/kg\ FO2} \tag{18}$$

$$\left[m_{SO_2}\right]_{LS} = 0.0161\ \mathrm{kg\ SO_2/kg\ LS} \tag{19}$$

The consumption of "*fuel N^er 2*" for conventional cycles was taking as a pattern, due to the fact that it was obtained in a rigorous way in the "Tests of Specific Consumptions and Costs" in all the Spanish Thermoelectric Power Plants, being the Department of Industry and Energy, across the DGESEN, the entity in charge of its application. This way, the value obtained considering a 370 MW full load electrical power in the above mentioned tests was:

$$\dot{m}_{FO2} = 86{,}145\ \mathrm{kg/h} \tag{20}$$

The aim now is to obtain an expression that allows getting the difference, in terms of consumption, between two different fuels (A) and (B) for the same net electrical power [50] as can be seen in eq. (21):

$$W_{N_{(A)}} = \frac{\eta_{(A)} \times \dot{m}_{fuel_{(A)}} \times LCV_{(A)}}{860} = W_{N_{(B)}} = \frac{\eta_{(B)} \times \dot{m}_{fuel_{(B)}} \times LCV_{(B)}}{860} \quad (21)$$

so finally eq. (22) can be obtained,

$$\dot{m}_{fuel_{(B)}} = \dot{m}_{fuel_{(A)}} \times \frac{\eta_{(A)}}{\eta_{(B)}} \times \frac{LCV_{(A)}}{LCV_{(B)}} \quad (22)$$

Values for the global efficiency [18] are: 38.41% for the *"fuel N^{er} 2"* and 38.72 % for the *"low sulphur fuel"*, so finally according to eq. (20):

$$\dot{m}_{LS} = 84{,}204 \text{ kg/h} \quad (23)$$

With these data, the hourly production of SO_2 both in mass and volume can be calculated as eqs. (24) and (25):

$$\left[\dot{V}_{SO_2}\right] = V_{SO_2} \times \dot{m}_{fuel} \quad (24)$$

$$\left[\dot{m}_{SO_2}\right] = m_{SO_2} \times \dot{m}_{fuel} \quad (25)$$

The theoretical results obtained in terms of emissions per hour for both fuels are shown in table 5 (first column). A considerable reduction in terms of emissions has been achieved by using *"low sulphur fuel"*.

Table 5. Theoretical and experimental values for the SO_2 volume flow.

SO_2	**$\dot{V}$ (Nm^3/h)**	
	Theoretical values	**Experimental results**
fuel N^{er} 2	1,757.35	1,865.43
low sulphur fuel	471.54	484.08

When using the unit mg/Nm^3, the standard volume (standard cubic meter, Nm^3) is used as a reference and the mass concentration of the flue gas is given in milligrams (mg). Since this unit is pressure and temperature dependent, the volume is expressed at standard conditions. There are different sets of standard conditions used for different purposes. Flue gas analysis commonly uses the following standard conditions: 0 °C and 1,013 mbar.

6.2. Experimental Results

Based on the data supplied by the portable, microprocessor-based gas analyzer, the combustion products of the power plant of reference burning both fuels are also summarised in table 5 (second column) in order to be able to compare results with the above mentioned theoretical ones. In this way a good agreement has been achieved.

7. Comparative Test: Conventional Cycles vs. Combined Cycles

7.1. Basic Principle and Further Improvements

The basic principle of combined cycles scheme is quite simple, in a combined cycle power plant (CCPP), a gas turbine alternator generates electricity and the waste heat from the gas turbine is used to make steam to generate additional electricity via a steam turbine; this last step enhances the efficiency of electricity generation. Most new gas power plants are of this type. In a thermal power plant, high-temperature heat as input to the power plant, usually from burning of fuel, is converted to electricity as one of the outputs and low-temperature heat as another output. As a rule, in order to achieve high efficiency, the temperature of the input heat should be as high as possible and the temperature of the output heat as low as possible (see Carnot efficiency). This is achieved by combining the Rankine (steam) and Brayton (gas) thermodynamic cycles. Figure 7 shows a simplified combined cycle scheme.

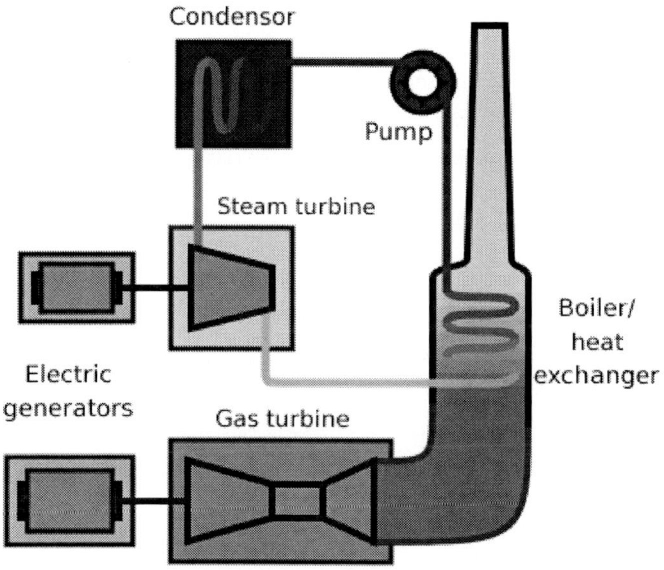

Figure 7. Working principle of a combined cycle power plant.

In a steam power plant water is the working medium. High pressure steam requires strong, bulky components. High temperatures require expensive alloys made from nickel or cobalt, rather than inexpensive steel. The high cost of these alloys limits practical steam temperatures to 655 °C. The lower temperature of a steam plant is fixed by the boiling point of water. As a result of this fixed maximum input and output temperature, a steam plant has a fixed upper efficiency of 35 to 40 %.

For gas turbines these limitations do not apply. Gas cycle firing temperatures above 1,200 °C are practicable. So, a combined cycle plant has a thermodynamic cycle that operates between the gas-turbine's high firing temperature and the waste heat is near the boiling point of water. A gas turbine has a compressor, burner and turbine. The input temperature to the turbine is relatively high (900 to 1,350 °C) but the output temperature of the flue gas is also high (450 to 650 °C).

The temperature of a gas turbine's flue gas is therefore high enough to make steam for a second steam cycle (a Rankine cycle), with a live steam temperature between 420 and 580 °C. The lowest temperature of the steam cycle depends on the boiling point of water, and is therefore fixed. The condenser is usually cooled by water from a lake, river, sea or cooling towers. The output heat of the gas turbine's flue gas is utilized to generate steam by passing it through a heat recovery steam generator (HRSG). By combining both processes, high input temperatures and low output temperatures can be achieved. The efficiency of the cycles adds, because they are powered by the same fuel source.

The number of this type of cycles installed is increasing around the world. It is also possible to use the steam from the boiler for heating purposes so such power plants can operate to deliver electricity alone or in combined heat and power mode [51]. Efficiencies are very wide ranging depending on the lay-out and size of the installation and vary from about 38 % for the conventional cycles to about 58 % for large new generation of combined cycle power stations. In the case of combined heat and power generation, the efficiency can increase to 85 %.

The HRSG can be designed with supplementary firing of fuel after the gas turbine in order to increase the quantity or temperature of the steam generated. Without supplementary firing, the efficiency of the combined cycle power plant is higher, but supplementary firing lets the plant respond to fluctuations of electrical load. Supplementary burners are also called "duct burners". More fuel is sometimes added to the turbine's exhaust. This is possible because the turbine exhaust gas (flue gas) still contains some oxygen. Temperature limits at the gas turbine inlet force the turbine to use excess air, above the optimal stochiometric ratio to burn the fuel. Often in gas turbine designs part of the compressed air flow bypasses the burner and is used to cool the turbine blades.

Another type of cycle is called: Integrated Gasification Combined Cycle, or IGCC, which is a power plant using synthetic gas (syngas). The gasification process can produce syngas from high-sulfur coal, heavy petroleum residues and biomass. The plant is called "integrated" because its syngas is produced in a gasification unit in the plant which has been optimized for the plant's combined cycle. The gasification process produces heat, and this is reclaimed by steam "waste heat boilers". The steam is utilized in steam turbines. There are currently (2007) only two IGCC plants generating power in the US; however, several new IGCC plants are expected to come online in the US. in the 2012-2020 time frames. There are several refinery-based IGCC plants in Europe that have demonstrated good availability (90-95 %) after initial shakedown periods.

In addition, it is a proven fact that when a boiler that has burned fuel oil for a long time is switched to natural gas, the deposits of H_2SO_4 (attached through the formation of acid dew point to the baskets and conduits on regenerative rotary air heaters) are pulverised and begin to dislodge after a few days [15-16]. The deposits of ash and clinker on the surfaces of the super-heater and re-heater are also dislodged; in other words, it has a self-cleaning effect on the boiler exhaust gases ducts.

Condensate of the combustion products for the natural gas presents initially an acid pH of about 3 (clearly validated in the daily practice of the power plant of reference). This is mainly due to the formation of carbonic acid but also to nitric acid.

On one hand, the carbonic acid, which is very volatile, is formed because of the closer contact existing between the water vapour and the carbon dioxide in the combustion exhaust gases, with a $K_a = 4.45\ 10^{-7}$. The formation route is:

$$CO_2 + H_2O \rightarrow H_2CO_3 \tag{26}$$

On the other hand, the formation of nitric acid is mainly due to the NO_X of the combustion exhaust gases. It is obtained by synthesis from the pentoxide of di-nitrogen, according to the following path [52]:

$$N_2O_5 + H_2O \rightarrow HNO_3 \tag{27}$$

Besides, in the combustion exhaust gases, the major quantity corresponds to nitric oxide (NO) and in minor quantity NO_2. The NO does oxidize rapidly and form NO_2 which also continues the oxidizing process to NO_3^- (acid aerosol) or directly to nitric acid, which can be dissociated itself completely in water. Then, the final pH value is of the order of 6 [53].

7.2. Analytical Calculation

An average value for LCV = 46,541.91 kJ/kg was assumed for natural gas [54-56]. Thus the NG consumption in conventional and combined cycles respectively, according to eq. (22), are:

$$[\dot{m}_{NG}]_{CON} = 86{,}145 \times \frac{38.41}{38.84} \times \frac{40{,}035.66}{46{,}541.91} = 73{,}282.10 \text{ kg NG / h} \tag{28}$$

$$[\dot{m}_{NG}]_{COM} = 86{,}145 \times \frac{38.41}{58.00} \times \frac{40{,}035.66}{46{,}541.91} = 49{,}073.74 \text{ kg NG / h} \tag{29}$$

Now, the amount of CO_2 emissions will be calculated next. First of all, the volumetric percentage composition of the natural gas provided by the gas chromatograph can be seen in table 6.

Table 6. Typical volumetric composition (%) of a NG international type, obtained from the PR.

Component (%)	**Maximum**	**Minimum**	**Average**
CH_4	89.8872	88.8317	88.9786
C_2H_6	5.1343	5.8767	5.1274
C_3H_8	1.2888	1.2662	1.2714
Iso-butane	0.2354	0.2249	0.2262
Normal-butane	0.3338	0.3283	0.3311
Iso-pentane	0.0794	0.0713	0.0727
Normal-pentane	0.0749	0.0718	0.0724
C_6H_{14}	0.1223	0.1177	0.1200
N_2	3.5393	3.4742	3.4898
CO_2	0.4185	0.2996	0.3112

The arrangement and grouping of the components of the matrix fuel with the corresponding combustion reactions considered can be seen in table 7.

Table 7. Combustion reactions for the different components of the *"natural gas"*.

Composition	**Basic combustion reactions (*)**
CH_4	$CH_4 + 2O_2 \rightarrow CO_2 + 2H_2O$
C_2H_6	$C_2H_6 + \frac{7}{2}O_2 \rightarrow 2CO_2 + 3H_2O$
C_3H_8	$C_3H_8 + 5O_2 \rightarrow 3CO_2 + 4H_2O$
C_4H_{10}	$C_4H_{10} + \frac{13}{2}O_2 \rightarrow 4CO_2 + 5H_2O$
C_5H_{12}	$C_5H_{12} + 8O_2 \rightarrow 5CO_2 + 6H_2O$
C_6H_{14}	$C_6H_{14} + \frac{19}{2}O_2 \rightarrow 6CO_2 + 7H_2O$

(*) The amount of CO was negligible, according to the data provided by the gas analyser.

The tests took place over three hours at stable load. During these tests the carbon monoxide reading provided by the gas analyser was insignificant and, therefore, no correction calculation is made and it is considered that carbon consumption produces solely carbon dioxide [57]. Thus, the number of moles of CO_2 in the combustion exhausts gases according to the reactions shown in table 7 is:

$$\psi_{CO_2} = 47.6181 \text{ mol } CO_2/Nm^3 \text{ NG} \tag{30}$$

So finally the volume and mass of CO_2, can be seen in eqs. (31) and (32) respectively.

$$\left[V_{CO_2}\right]_{NG} = \psi_{CO_2} \times \frac{22.412}{1{,}000} = 1.0666\ Nm^3 CO_2 / Nm^3\ NG \tag{31}$$

$$\left[m_{CO_2}\right]_{NG} = \psi_{CO_2} \times \frac{M_{CO_2}}{1{,}000} = 2.0956\ kg\ CO_2 / kg\ NG \tag{32}$$

According to eq (32) and eqs. (28) and (29) respectively the volume flow of CO_2 emissions for both conventional and combined cycles are:

$$\left[\dot{m}_{CO_2}\right]_{CON} = \left[m_{CO_2}\right]_{NG} \times \left[\dot{m}_{NG}\right]_{CON} = 216{,}621.88\ kg\ CO_2 / h \tag{33}$$

$$\left[\dot{m}_{CO_2}\right]_{COM} = \left[m_{CO_2}\right]_{NG} \times \left[\dot{m}_{NG}\right]_{COM} = 102{,}838.93\ kg\ CO_2 / h \tag{34}$$

As has been demonstrated, a great amount of CO_2 emissions savings have been achieved using combined cycles instead of the conventional ones schemes.

8. European Directives Regarding Emissions in Power Plants

8.1. Aim and Goals

European Commission activities to protect the air concern a wide range of problems: limiting depletion of stratospheric ozone, controlling acidification, ground-level ozone and other pollutants and climate change.

The EC has taken important steps over the past decade, but there is no substantial improvement in the environment yet, in spite of a decrease in the emissions to air and water of a number of pollutants, such as sulphur dioxide (a 50 % reduction since 1980), lead (a 60 % reduction since 1980), phosphorous in many water catchments' areas (a 30 to 60 % reduction since 1980) and to a lesser extent nitrogen oxides and volatile organic compounds (a 14% reduction since 1990). Atmospheric pollutants, which enter the air from a wide variety of sources, can be subdivided into three broad categories:

- Emissions from mobile sources (transport industry). Apart from CO_2, the main ones are: nitrogen oxides (NOx), carbon monoxide (CO) and hydrocarbons (HC), i.e. volatile or non-volatile organic compounds, soot particles and ozone (O_3).
- Emissions from immobile source (businesses, homes, farms and rubbish dumps). Apart from CO_2, the main ones are: sulphur dioxide (SO_2), nitrogen oxides (NOx), hydrocarbons (HC), soot particles, chlorofluorocarbons (CFCs) and methane.
- Emissions caused by power generation. Apart from CO_2, the main ones are sulphur dioxide (SO_2) and soot particles.

High concentrations of these gases and pollutants arising from them through chemical reactions in the atmosphere or in the soil are harmful to human health, corrode various materials and damage vegetation, have a detrimental effect on agricultural and forestry production and cause unpleasant smells. Many of these pollutants, such as carbon dioxide (CO_2), methane, nitrogen oxides (NOx) and chlorofluorocarbons (CFCs), are responsible for the greenhouse effect.

8.2. European Commission's Proposal

To reduce sulphur dioxide (SO_2) emissions the EC adopted a number of measures, the most important of which is the directive on large combustion plants, 88/609, as last amended by Directive 94/66. Taking 1980 as the reference year it aims to reduce annual emissions of SO_2 by 73 % by the year 2000. The EC directive on limit values for SO_2 emissions requires a reduction of 57 % from 14.4 in 1980 to 6.2 million tonnes by the year 2005.

The Commission proposed a framework directive in (COM (97)0500) to set limit values for sulphur dioxide, nitrogen oxides, particulate matter and lead in ambient air, which the political authorities are currently discussing. In July 1998 the Commission put forward a new proposal for a Directive amending Council Directive 88/609 on the limitation of emissions of certain pollutants into the air from large combustion plants (COM(98)0415).

Countries of the European Union are subject to its Directive 1999/32/EC requiring them to reduce sulphur in heavy fuel oil (HFO) to 1 % (wt.) and sulphur in light fuel oil (LFO) to 0.1 % (wt.) by January 1, 2008. Some countries such as Austria, Denmark and Finland already have in place a limit of 1% (wt.) sulphur (or less) for HFO. The current EU sulphur limit in LFO is a maximum of 0.2 % (wt.) Table 8 shows the amending acts to the directive 93/12/EEC for verification of the sulphur content of marine fuels. Every year, Member States must send the Commission a report on the sulphur content of the fuels covered by this proposal and used on their territory. By 31 December 2010, the Commission must send Parliament and the Council a report on implementation of the Directive, together with any proposals for amending it.

Table 8. Amending acts to the directive 93/12/EEC for verification of the sulphur content.

Amending act(s)	Entry into force	Deadline in the Member States	Official Journal
98/70/EC	28.12.1998	1.7.1999	OJ L 350, 28.12.1998
99/32/EC	11.5.1999	1.7.2000	OJ L 121, 11.5.1999
2005/33/EC	11.8.2005	11.8.2006	OJ L 191, 22.7.2005

This proposal is intended, inter alia, to confirm 1 January 2009 as the date on which all diesel must have a maximum sulphur content of 10 ppm (parts per million); to reduce the polyaromatic hydrocarbon content of diesel to 8 % (instead of 11 %); to reduce the maximum sulphur content of non-road gas-oil from 1,000 ppm to 10 ppm for land-based uses and from

1,000 ppm to 300 ppm for inland waterways; to increase the oxygenate content and vapour pressure limit for petrol blended with ethanol, in order to enable a higher volume of biofuels to be used in petrol, with appropriate labelling; and to require fuel suppliers to reduce life-cycle (refining, transport and use) greenhouse gas emissions from these fuels by 1 % per year from 2009.

Its objective is to reduce emissions of SO_2 across the European Union. According to the European Commission, it "is an integral part of a cost-effective strategy designed to combat acidification as well as reducing atmospheric pollution by sulphur dioxide and particulate matter".

Directive 2001/80/EC on the limitation of emissions of certain pollutants into the air from Large Combustion Plants and Directive 2001/81/EC on national emissions ceilings for certain atmospheric pollutants, form part of an overall European strategy to combat acidification. In November 2000, the European Union Council adopted national emission ceiling limits (to be met by 2010) for certain atmospheric pollutants including SO_2 [58]. Accordingly, some flexibility is built in the Directive to "reflect the conclusions from the integrated assessment (for the acidification strategy) and to avoid non cost-effective expenditure". Table 9 shows the national emission ceilings for SO_2, NO_x and VOC (volatile organic compounds), to be achieved by 2010 for some of the EU countries [59].

Table 9. Some national emission ceilings (kiloton's) for SO_2, NO_x and VOC, to be achieved by 2010.

Country	SO_2	NO_X	VOC
France	375	810	1,050
Germany	520	1,051	995
Italy	475	990	1,159
Luxembourg	4	11	9
Netherlands	50	260	185
Portugal	160	250	180
Spain	746	847	662
Sweden	67	148	241
United Kingdom	585	1,167	1,200

Thus, the EU Directive includes the following flexibility:

1. The 1% (wt.) sulphur limit for HFO does not apply for fuel used in combustion plants that are:
   a. equipped with effective desulphurisation technologies (as per Directive 88/609/EEC on large combustion plants);
   b. have thermal input below 50 megawatts where the emissions of sulphur dioxide are at concentrations less than or equal to 1,700 mg/Nm^3 (in flue gas

with oxygen content 3 % by volume on dry basis at standard temperature and pressure);

   c. or are part of refineries where monthly average emission limits do not exceed 1,700 mg/Nm^3.

2. The requirement for sulphur in HFO does not apply to fuels for marine use.
3. While the sulphur requirement for light fuel oils applies to marine gas oils (i.e. distillate fuels intended for marine use), it allows for a derogation for these fuels in Greece, the Spanish Canary Islands, the French Overseas Departments, and the Portuguese archipelago of Madeira and Azores, since low sulphur levels "may present technical and economic problems" for these countries.
4. The EU Directive allows a member state to authorize HFO with sulphur levels between 1 % (wt.) to 3% (wt.) to be used in part or the whole of its territory if air quality standards for SO_2 are respected and where emissions do not contribute significantly to acidification in any member state. The Directive specifies that to avail itself of this possibility, a member state has to inform the European Commission and the public at least 12 months beforehand. The Directive establishes a process where proposed measures are reviewed by the Commission which then makes a decision. According to the Directive, the Commission's decision is to be reviewed every eight years on the basis of information to be provided to the Commission by member states.
5. The Directive allows a member state to authorize, up to the end of 2012, LFO with sulphur levels between 0.1 % (wt.) to 0.2 % (wt.) to be used in part or the whole of its territory if air quality standards for SO_2 are respected and where emissions do not contribute significantly to acidification in any member state. To avail itself of this possibility, a member state would follow the process identified above for HFO.

According to the Commission, these flexibilities recognize that for power stations and certain industries it would be more cost effective to remove SO_2 from emissions using technology such as flue gas desulphurisation rather than using low sulphur HFO. The Commission considers that the emission standard of 1,700 mg SO_2/Nm^3 "is approximately the equivalent, in terms of emissions, of burning heavy fuel oil with a 1 % (wt.) sulphur content".

The European Commission estimated that the introduction of the 1% (wt.) sulphur limit for HFO "will reduce SO_2 emissions in 2010 by approximately 1 million tonnes as compared to what would be the case in the absence of the Commission's proposal". In considering benefits to human health, the Commission estimated that, on average, the overall economic cost of the damage resulting from 1 million tonnes of SO_2 emissions to be approximately 4 billion EUROS [60].

Based on the provisions of the Directive, Member States are obliged to report each year their national emission inventories and projections for 2010 to the European Commission and the European Environment Agency. They shall also draw up national programs twice in order to demonstrate how they are going to meet the national emission ceilings by 2010. The first year for drawing up national programmes was 2002. Member States are obliged to send their second national programme to the Commission by the end of 2006.The Directive provides for review in 2004 and 2008 with a view to making up the shortfall.

9. Implementing a Practical Method for Reducing Acid Emissions

9.1. Setting Up the Context

There is the legal requirement to remain below a certain level of pollutant output and to operate the equipment inside the safety limits at all times. Failure to adhere to these points may well result in heavy fines or the closure of all the equipment until service has been carried out and it can be proved that the safety limits are no longer exceeded or until EPA (or any other Environmental Agencies) compliance testing has been repeated successfully. These procedures are always costly and may lead to long delays and shutdown of a plant, possibly at a very inopportune moment. The safety aspect will, of course, be clear to anyone who has had discussions with the local safety at work or equivalent officials. Safety is important for many reasons, not just financial ones.

This study is focused on an experimental way to reduce acid emissions in the power plant of reference working at minimum technical load (130 MW), studying different combinations of using of fuel oil burners and natural gas igniters. Thus, this comparative test addresses the optimisation of the operation mode for a thermal power plant using the best of these specific combinations, analysing this way the economic and environmental advantages of the finally proposed solution.

As described previously in the introduction section, "alternative A" or traditional mode, consisted in operating the boiler by firing the 7 pairs of low-sulphur fuel oil burners as well as the 2 pairs of natural gas igniters in order to ensure the burners start-up, shut-down and venting procedures respectively. Thus, the procedure proposed in this study, so called "alternative B" consist of operating the boiler by firing up just 6 pairs of low-sulphur fuel oil burners, plus the whole of the 12 pairs of natural gas igniters available in the plant, which besides ensuring the operating safety and requirements of the low sulphur fuel burners would operate now as real natural gas burners.

The most significant values of the various operating modes studied were obtained on the same date with fully stable loads maintained in each test for regular 3-hour periods. The analytical calculation procedure is shown for the respective SO_2 and CO_2 emissions in order to provide finally the different operating costs associated with the two methods, so the latest Spanish market prices could be applied accordingly.

The main output operating data obtained for the two alternatives are presented in Table 10.

Table 10. Fuel consumption for alternatives A and B at minimum technical load.

	Alternative "A"	Alternative "B"
$\dot{m}_{LS}$ (kg/h)	34,368	26,856
$\dot{V}_{NG}$ (Nm^3/h)	1,400	9,005

The adoption of the alternative proposed has of course no negative effect on steam temperatures, which are even slightly increased by the use of a larger quantity of natural gas corresponding to alternative B.

9.2. Comparative of SO_2 Emissions

According to available data from table 10, the emissions of SO_2 can be easily obtained for both alternatives, such as:

$$\left[\dot{V}_{SO_2}\right]_A = 0.0056 \cdot 34{,}368 = 192.461 \text{ Nm}^3/\text{h} \tag{35}$$

$$\left[\dot{V}_{SO_2}\right]_B = 0.0056 \cdot 26{,}856 = 150.393 \text{ Nm}^3/\text{h} \tag{36}$$

So a substantial reduction has been achieved with this alternative proposed as only "*low sulphur fuel*" is the source for these emissions.

9.3. Comparative of CO_2 Emissions

- Burning "Low Sulphur Fuel"

The amount of CO_2 emissions burning "*low sulphur fuel*" will be calculated such as:

$$\left[\psi_{CO_2}\right]_{LS} = \psi_C = 71.5903 \text{ mol } CO_2 / \text{kg LS} \tag{37}$$

And therefore the volume and mass of CO_2 (see eqs. (38) and (39) respectively):

$$\left[V_{CO_2}\right]_{LS} = \psi_{CO_2} \times \frac{22.412}{1{,}000} = 1.5936 \text{ Nm}^3 CO_2 / \text{kg LS} \tag{38}$$

$$\left[m_{CO_2}\right]_{LS} = \psi_{CO_2} \times \frac{M_{CO_2}}{1{,}000} = 3.1506 \text{ kg } CO_2 / \text{kg LS} \tag{39}$$

According to available data from table 10, the emissions of CO_2 from the "low sulphur fuel" can be easily obtained for both alternatives, such as:

$$\left[\dot{V}_{CO_2}\right]_{A,LS} = 1.5936 \times 34{,}368 = 54{,}768.84 \text{ Nm}^3 \text{ } CO_2 / \text{h} \tag{40}$$

$$\left[\dot{V}_{CO_2}\right]_{B,LS} = 1.5936 \times 26{,}856 = 42{,}797.72 \text{ Nm}^3 \text{ } CO_2 / \text{h} \tag{41}$$

and the same can be said in terms of mass, so:

$$\left[\dot{m}_{CO_2}\right]_{A,LS} = 3.1506 \cdot 34{,}368 = 108{,}282.88 \text{ kg } CO_2 / h \quad (42)$$

$$\left[\dot{m}_{CO_2}\right]_{B,LS} = 3.1506 \cdot 26{,}856 = 84{,}614.90 \text{ kg } CO_2 / h \quad (43)$$

- Burning "Natural Gas"

Also according to available data from table 10, the emissions of CO_2 from the "*natural gas*" can be easily obtained for both alternatives, such as:

$$\left[\dot{V}_{CO_2}\right]_{A,NG} = 1.0666 \cdot 1{,}400 = 1{,}423.24 \text{ Nm}^3 \; CO_2 / h \quad (44)$$

$$\left[\dot{V}_{CO_2}\right]_{B,NG} = 1.0666 \cdot 9{,}005 = 9{,}604.73 \text{ Nm}^3 \; CO_2 / h \quad (45)$$

and the same again can be said in terms of mass, so:

$$\left[\dot{m}_{CO_2}\right]_{A,NG} = 2.0956 \cdot 1{,}400 = 2{,}933.94 \text{ kg } CO_2 / h \quad (46)$$

$$\left[\dot{m}_{CO_2}\right]_{B,NG} = 2.0956 \cdot 9{,}005 = 18{,}871.53 \text{ kg } CO_2 / h \quad (47)$$

Comparative global results for both alternatives are depicted in figure 8 where important improvements in relation to the environment have been achieved with the alternative proposed here especially in relation to CO_2 emissions.

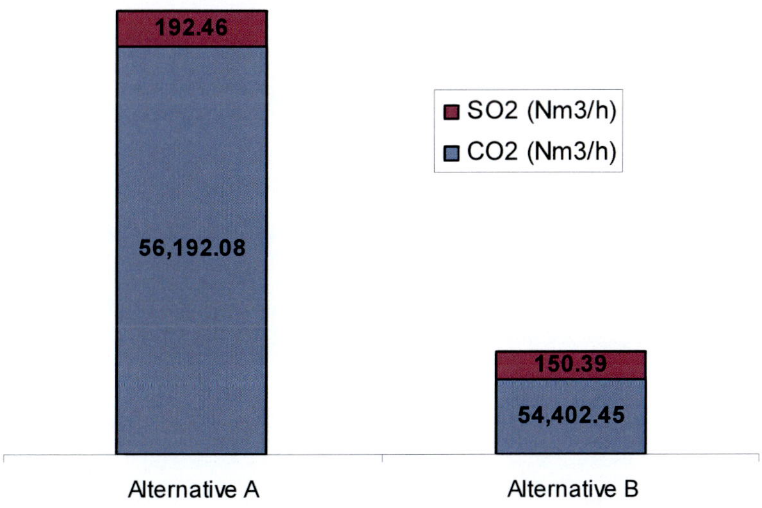

Figure 8. Comparative environmental results (CO_2 and SO_2) for alternatives A and B.

9.4. Experimental Results

Finally, table 11 shows other environmental results (emissions of CO, SO_2 and NO_x) obtained from experimental test carried out with the gas analyzer for both alternatives at the same working conditions of the power plant, so it can be seen that savings are not negligible at all showing good agreement with the analytical results. The amount of CO is negligible as has been previously exposed.

Table 11. Experimental results obtained from the gas analyzer for both alternatives.

(mg/Nm^3)	Alternative "A"	Alternative "B"
CO	1.3	1.0
SO_2	1,248	995
NO_x	596	547
Ω (%)	8.9	6.9

9.5. Comparative of the Cycle Efficiency

The cycle efficiency can be defined such as:

$$\eta_c = \frac{\dot{Q}_{b\to c} + \dot{Q}_{b\to a}}{\dot{W}_N} \tag{48}$$

where:

$$\dot{Q}_{b\to c} = (\dot{m}_{vp} \times i_{vp}) + (\dot{m}_{vRC} \times i_{vRC}) - (\dot{m}_{aa} \times i_{aa}) - (\dot{m}_{vRF} \times i_{vRF}) - (\dot{m}_{atR} \times i_{atR}) \tag{49}$$

$$\dot{Q}_{b\to a} = (\dot{m}_{vs} \times i_{vs}) + (\dot{m}_{pc} \times i_{pc}) - (\dot{m}_{va} \times i_{va}) - (\dot{m}_{vRF} \times i_{vRF}) - (\dot{m}_{vf} \times i_{va}) \tag{50}$$

Table 12. Increase of the cycle efficiency between both alternatives.

	Alternative "A" ($x\ 10^6$)	Alternative "B" ($x\ 10^6$)
$\dot{Q}_{b\to c}$ (kcal/h)	1,078.939	1,107.642
$\dot{Q}_{b\to a}$ (kcal/h)	18.783	18.783
$\Delta\eta_c$ (%)	2.55	

Table 12 summarises the main comparative results of the two alternatives with reference to cycle efficiency. So the increase of the cycle efficiency is an important improvement derived from this change of operation mode.

9.6. Financial Cost for Both Alternatives

At present, prices of both fuels per "therm" produced (1 thm = 105.506 10^3 kJ) [61] and for CO_2 emissions per tonne emitted into the atmosphere respectively, (according to the Spanish current legislation) were as follows:

- Low-sulphur fuel: 0.02809 € / thm. (51)

- Natural gas: 0.03037 € / thm. (52)

- CO_2: 16.15 € / tonne. (53)

Be observed that the prices of fuel are invoiced in Spain on the base of the lower calorific value (LCV) which has a clear effect with regard to the yield of boiler, since the combustion of the natural gas means a yield increasing of the order of 1 %, whereas if the invoicing was on the base of the gross calorific value, a loss of yield would take place in boiler of the order of 3 % [62]. With these prices, the respective costs for the two alternatives can be calculated as follows:

- Alternative A:

The financial cost of burning low-sulphur fuel oil for this alternative is:

$$(C)_{LS,A} = 2.662 \cdot 10^{-7}\ \mathrm{Eur/kJ} \times 40{,}630.68\ \mathrm{kJ/kg} \times 34{,}368\ \mathrm{kg/h} = 371.720\ \mathrm{Eur/h} \quad (54)$$

The financial cost of burning natural gas for this alternative is:

$$(C)_{NG,A} = 2.87 \cdot 10^{-7}\ \mathrm{Eur/kJ} \times 37{,}426.8\ \mathrm{kJ/Nm^3} \times 1{,}400\mathrm{Nm^3/h} = 15.043\ \mathrm{Eur/h} \quad (55)$$

The financial cost associated with the production of CO_2 for this alternative is:

$$(C)_{CO_2,A} = 16.15 \cdot 10^{-3}\ \mathrm{Eur/kg} \times 111{,}216.82\ \mathrm{kg/h} = 1{,}796.15\ \mathrm{Eur/h} \quad (56)$$

- Alternative B:

Following the same procedure, the financial cost of burning low-sulphur fuel oil and natural gas and the financial cost associated with the production of CO_2 for this alternative can be calculated, so figure 9 presents a comparative overview of the financial results obtained for this alternative.

As may be deduced from this figure, the financial saving in overall cost operating with "alternative B" as opposed to "alternative A" is about 124 €/h at minimum technical load which is not negligible at all.

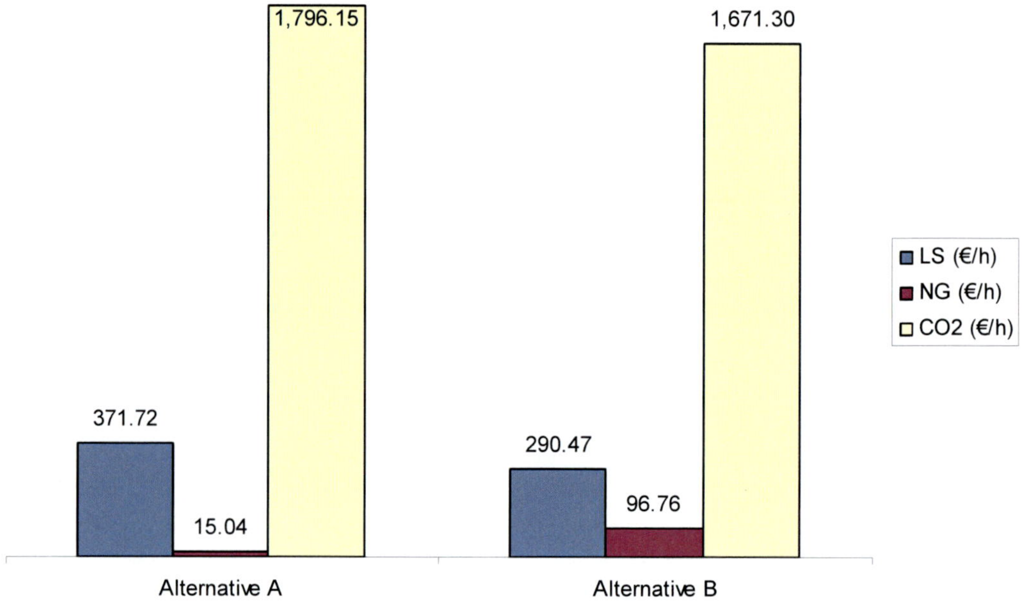

Figure 9. Partial and total cost estimated for both alternatives.

9.7. Considerations about Fluctuations in Fuel Prices

Crude oil prices behave much as any other commodity with wide price swings in times of shortage or oversupply. The crude oil price cycle may extend over several years responding to changes in demand as well as OPEC and non-OPEC supply.

The US petroleum industry's price has been heavily regulated through production or price controls throughout much of the twentieth century. In the post World War II era US oil prices at the wellhead have averaged $23.57 per barrel adjusted for inflation to 2006 dollars. In the absence of price controls the US price would have tracked the world price averaging $25.56. Over the same post war period the median for the domestic and the adjusted world price of crude oil was $18.43 in 2006 prices. That means that only fifty percent of the time from 1947 to 2006 have oil prices exceeded $18.43 per barrel.

Until March the 28th, 2000 adoption of the $22-$28 price band for the OPEC basket of crude, oil prices only exceeded $23.00 per barrel in response to war or conflict in the Middle East. With limited spare production capacity OPEC has abandoned its price band and for close to three years was powerless to stem a surge in oil prices which was reminiscent of the late 1970's. Figure 10 shows the fluctuations of crude prices since 1947, according to the most important political and economical events happened since then.

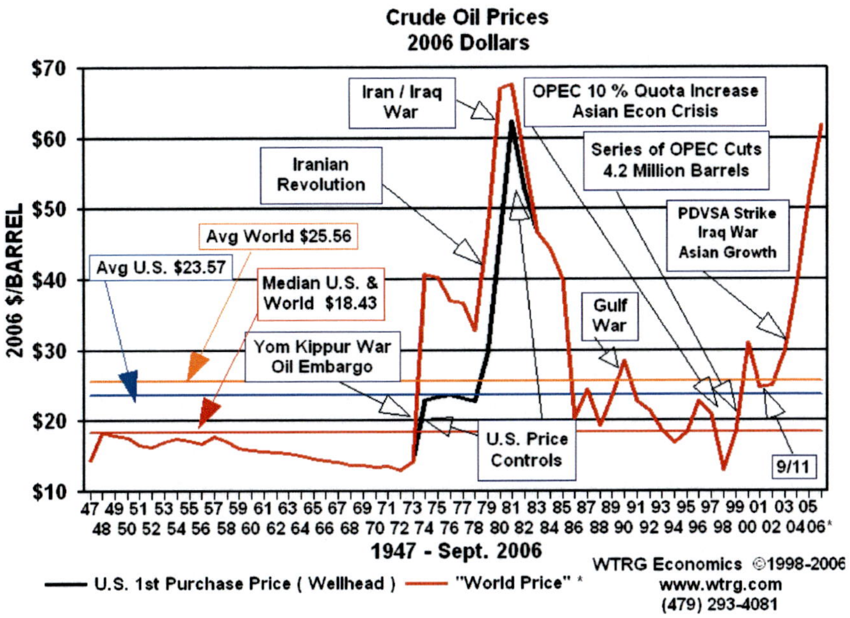

Figure 10. Fluctuations of crude prices since 1947 (courtesy of WTRG economics; http://www.wtrg.com).

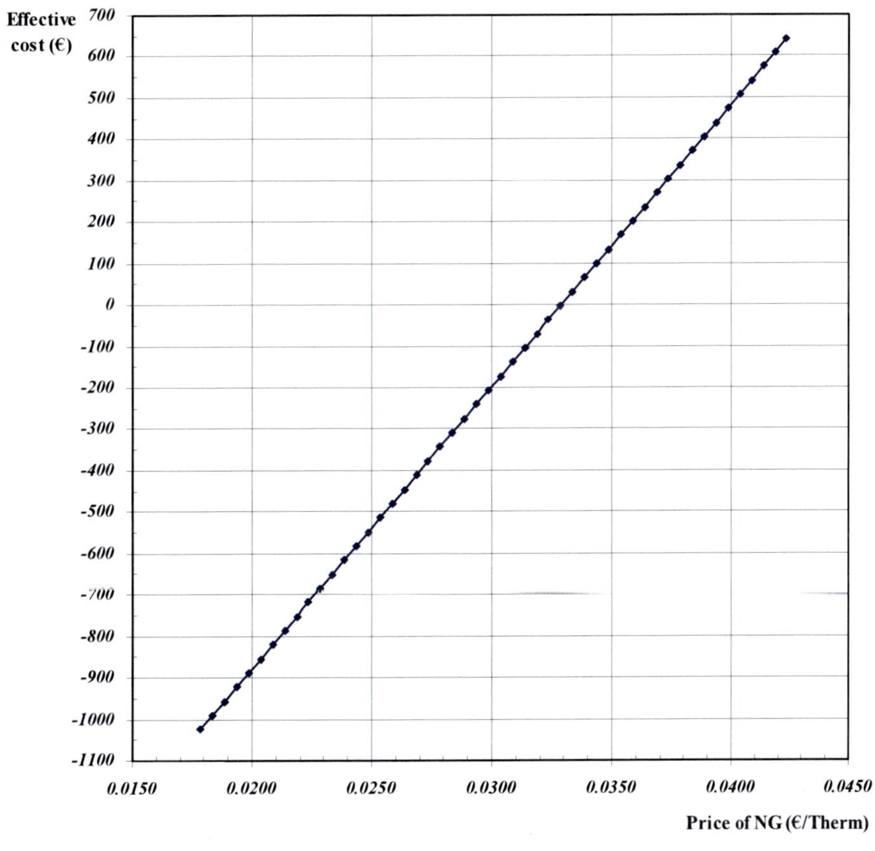

Figure 11. Variation of the natural gas price, keeping constant the low sulphur fuel oil price (0.02809 €/thm.)

The effect of the fluctuations in the price of fuels for the installation to be profitable will be analyzed next. Thus a spreadsheet containing different combinations of prices for the low-sulphur fuel oil and natural gas has been performed for "alternative B". In this way, keeping constant the price of one of the items and varying the price of the other one can establish the profitable combinations. The price of CO_2 has been kept constant as its variation is not so relevant.

On the one hand, figure 11 shows the effective cost (which includes the costs for low sulphur fuel, natural gas and CO_2 respectively) for different combinations of prices, where the low sulphur fuel price is kept constant with the value of reference (0.02809 €/thm) while natural gas price shows a wide variation range around its reference value of 0.03037 €/thm.

Profitable combinations have been defined here by negative values for the effective cost which means that with the present price (reference value) of the low-sulphur fuel oil, the price of natural gas could increase from its reference value up to 0.03287 €/thm in order to be considered as a profitable combination. Prices for the natural gas below this reference value would correspond of course to profitable combinations although this situation will be very difficult to get nowadays according to the recent market trends.

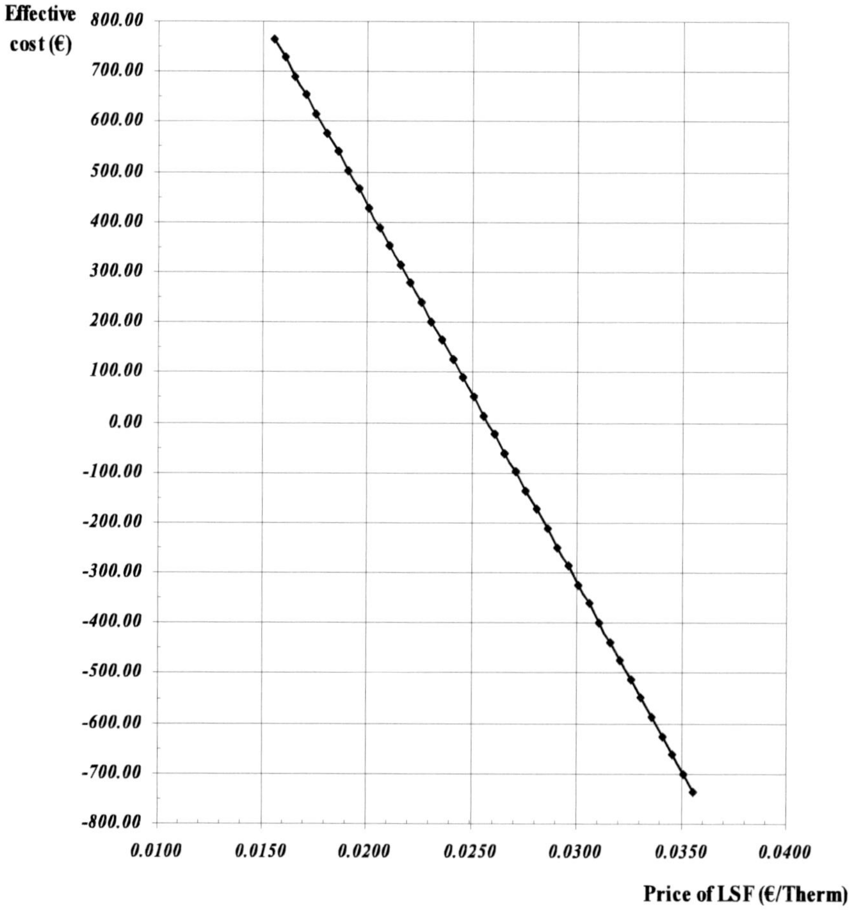

Figure 12. Variation of the low sulphur fuel oil price, keeping constant the natural gas price (0.03037 €/thm.)

On the other hand, figure 12 also shows the effective cost but in this case for situations where the variation range corresponds to the price of the low-sulphur fuel oil, while the price of natural gas is kept constant with its reference value (0.03037 €/thm).

10. Conclusion

According to this work, a semi experimental-analytic method has been showed for determining the Acid Dew point Temperature and thus, time affected by acidic corrosion inside air preheaters in thermo-electric power plants through the use of data obtained in a power plant of reference depending of the fuel burned.

An in-depth thermal analysis was previously carried out on the transient process of heat exchange within a rotating regenerator, and a time dependent surface heating-cooling curve was obtained that will serve as a basis for determining the time per cycle and sector of baskets potentially exposed to ADT, depending also on the type of fuel burned.

Bearing in mind that the manufacturer of the rotating regenerators recommends checking and replacing baskets every 4 years when burning "*fuel N^{er} 2*", the fact of replacing this fuel with "*low sulphur fuel*" would extend the maintenance time by around 1 year more, thereby leading to a considerable extra financial saving.

We should not ignore the environmental impact. As far as SO_2 emissions, conventional cycles burning *"low sulphur fuel"* save almost 73 % with regard to the same cycle scheme, burning "*fuel N^{er} 2*".

Experimental values obtained about ADT over the thermal power plant of reference, are very close to the theoretical values given in this paper, so this methodology has been revealed as a very good approach.

As far as SO_2 emissions, conventional cycles burning *"low sulphur fuel"* save almost 73 % with regard to the same cycle scheme, burning *"fuel oil N^{er}2"*.

Combined cycles burning natural gas, save practically 33 % of CO_2 emissions with regard to conventional cycles.

The use of natural gas igniters as support of low-sulphur fuel oil burners for thermal power plants operating at minimum technical load (130 MW) in a combination proposed here (alternative B) different from the traditional mode of operation (alternative A) has revealed as a good procedure. This proposed operating method therefore consist to operate the boiler by firing 6 pairs of low-sulphur fuel oil burners, plus 12 pairs of natural gas igniters, which besides ensuring the operating safety of the burners would also operate as proper burners of natural gas.

Nowadays not only the price of the fuel but also CO_2 emissions should be considered as a factor in the production of thermal energy. The operating mode proposed (alternative B) yields in a clear reduction, in terms of emissions, as compared to the traditional mode (alternative A). Thus, the following improvements were achieved:

- Financial gain of about €124 per hour of operation at minimum technical load.
- Significant reduction of around 7 % in the CO_2 emissions in terms of mass.
- About 8 % reductions in NO_x emissions in terms of mass (experimental).
- About 20 % reductions in SO_2 emissions in terms of mass.
- 2 % reduction in Ω (experimental).

- 2.55 % increase of the cycle efficiency (η_c).

Fluctuations in prices of fuels are of a great significance as have been showed here, where several cases have been studied in order to determine profitable combinations of prices, allowing a range of variation of prices for both fuels studied around the current ones. Thus, while the price of natural gas could increase in about 0.0025 €/thm., the price of low-sulphur fuel oil could increase more than 0.0075 €/thm.

Finally, other experimental considerations concluded through this work, are:

When burning natural gas, the outlet temperature on the furnace tends to rise and this in turn tends to increase temperatures [6] and, therefore, the enthalpy of the superheated and reheated steam, bringing them closer to rated design specifications, with the ensuing improvement in cycle performance [63]. The increase in temperatures that would be obtained by applying the proposed operating mode does not compromise the integrity of the superheater at all.

In addition, it is a proven fact that when a boiler that has burned fuel oil for a long time is switched to natural gas, the deposits of H_2SO_4 (attached through the formation of acid dew point to the baskets and conduits on regenerative rotary air heaters) are pulverised and begin to dislodge after a few days. The deposits of ash and clinker on the surfaces of the superheater and re-heater are also dislodged; in other words, it has a self-cleaning effect on the boiler exhaust gases ducts.

Acknowledgements

Authors are deeply thankful to all staff of the power plant of reference for their help and support on the use of the related facilities for the test experiments carried out in this work.

Nomenclature

a	=	Thermal diffusivity = $\left(\frac{\lambda}{\rho \cdot c_p}\right)$ (m^2/s)
ADT	=	Acid Dew point temperature (°C).
c_p	=	Specific heat at constant pressure (kJ/kg K).
C	=	Cost (€ / h)
CCPP	=	Combined Cycle Power Plant.
CEN	=	Fuel net specific consumption.
CO	=	Carbon monoxide.
CO_2	=	Carbon dioxide.
DG	=	Dry gases.
EPA	=	Environmental Protection Agency.
EC	=	European Commission.
EU	=	European Union.
HFO	=	Heavy fuel oil.
HRSG	=	Heat Recovery Steam Generator.

H_2SO_3 = sulphurous acid.
H_2SO_4 = Sulphuric acid.
i = Specific net enthalpy (kcal/kg h).
IGCC = Integrated Gasification Combined Cycle.
K_a = Acidity constant.
LCD = Liquid Cristal Display.
LCV = Lower calorific value (kJ/ kg fuel).
LFO = Light fuel oil.
m = Mass (g).
$\dot{m}$ = Mass flow (kg/h).
M = Molecular weight (g/mol).
NECD = National Emission Ceilings Directive.
NO_x = Nitrogen oxides.
OPEC = Organization of the Petroleum Exporting Countries.
ppm = Parts per million.
PR = Power plant of reference.
$\dot{Q}_{b \to c}$ = Heat power yielded by the boiler to the cycle (kJ/h).
$\dot{Q}_{b \to a}$ = Heat power yielded by the boiler to the auxiliary systems (kJ/h).
SO_2 = Sulphur dioxide.
$[SO_2]_m$ = Sulphur dioxide mass concentration (mg/Nm^3).
$[SO_2]_V$ = Sulphur dioxide volume concentration (ppm).
SO_3 = Sulphur trioxide.
SO_x = Sulphur oxides.
t = Time (s).
T = Wave period (s^{-1}).
US = United States (of America).
V = Volume (Nm^3).
VOC = Volatile organic compounds.
$\dot{V}$ = Volume flow (Nm^3/h).
W_N = Net power (kWe).
x = Distance from surface (mm).

Greek symbols

α = *Total rotation angle of the rotating regenerator (º).*

α_1, α_2 = *Sectors of baskets directly affected by ADT for* "fuel N^{er} 2" *and* "low sulphur fuel" *(º).*

Δ = *Incremental.*

π = *Number Pi (3.141516).*

θ = *Temperature (ºC).*

λ = *Thermal conductivity of the baskets (W/m K).*

ψ	=	*Number of moles.*
∂	=	*Derivative.*
ρ	=	*Density of the panels (kg/m^3).*
η_c	=	*Cycle efficiency.*
η_N	=	*Global net efficiency (= 1/CEN).*
Ω	=	*Gases opacity (%).*

Subscripts

(a)	=	*Relative to air.*
(aa)	=	Relative to the reheated vapour chill water.
(atR)	=	Relative to the boiler feed water.
(A)	=	*Relative t alternative A.*
(av)	=	*Relative to the average value.*
(B)	=	*Relative to alternative B.*
(c e)	=	*Relative to the cold end temperature.*
(COM)	=	*Relative to combined cycles.*
(CON)	=	*Relative to conventional cycles.*
(DG)	=	*Relative to dry gases.*
(FO2)	=	Relative to fuel oil number 2.
(fuel)	=	Relative to the fuel.
(g)	=	*Relative to the exhaust gases.*
(h e)	=	*Relative to the hot end temperature.*
(i)	=	*Relative to the inlet.*
(LS)	=	Relative to low sulphur fuel.
(m)	=	*Relative to mass.*
(max)	=	*Relative to a maximum value.*
(NG)	=	*Relative to natural gas.*
(o)	=	*Relative to the outlet.*
(s)	=	*Relative to the surface.*
(V)	=	*Relative to volume.*
(vp)	=	*Relative to the main vapour flue.*

(vRC) = *Relative to the reheated vapour flue (hot).*

(vRF) = *Relative to the reheated vapour flue (cold).*

(wt) = *Relative to terms of weight.*

References

[1] Beér, J.M., 2000, Combustion technology developments in power generation in response to environmental challenges. *Prog. Energy Combust Sci.*, **26**, p. 301 - 27.

[2] Weber, R., Verlaan, A.L., Orsino, S. and Lallemant, N., 1999, On emerging furnace design methodology that provides substantial energy savings and drastic reductions in CO_2, CO and NO_x emissions. *J Inst Energy* **72** 492, p. 77 - 83.

[3] Barnes, F.J., Bromly, J.H., Edwards, T.J. and Madngezewsky, R., 1988, NO_x Emissions from Radiant Gas Burners. *Journal of the Institute of Energy*, **155**, p. 184 - 88.

[4] Chacón, J., Sala, J.M. and Blanco, J.M., 2007, Investigation on the Design and Optimization of a Low NOx-CO Emission Burner Both Experimentally and through CFD simulations, *Energy and Fuels*. **21** (1), p. 42 - 58.

[5] Fenimore, C.P., 1971, Formation of nitric oxide in premixed hydrocarbon flames. *13th Symposium (International) on Combustion*, The Combustion Institute, Pittsburgh (USA), p. 373 - 380

[6] Chagger, H.K., Goddard, P.R., Murdoch, P. and Williams, A., 1991, Effect of SO_2 on the reduction of NO_X by reburning with methane, *Fuel*, **70**, p. 1137 - 42.

[7] Rodhe, H., et. Al., 1998, The Global Distribution of Acidifying Wet Deposition, *Environmental Science & Technology*, **36**, 20, p. 4382-8.

[8] Mc. Cormick, J., Acid Earth: The Global Threat of Acid Pollution, London: Earthscan, 1989.

[9] http://www.epa.gov/acidrain.

[10] Corley, T.L., Wendt, J.O.L., 1984, Post flame behaviour of nitrogenous species in the presence of fuel sulphur: II. Rich, $CH_4/He/O_2$ flames. *Combustion and Flame*, **58**, p. 141 - 52.

[11] Cunnignham, A.T.S., and Datschefski, G., 1981, *The effects of heavy fuel oil composition on particulate emissions from power station boilers*. The Heavy End of the Barrel-Future Trends in Oil Firing, Institute of Energy Symposium, The Institute of Energy, Portsmouth (England), p. 126 - 46.

[12] Burgess, A.R. and Molero, L.J., 1996, *Combustion generated NO_X and SO_2 from heavy liquid fuels.* Joint Meeting of the Portuguese, British, Spanish and Swedish Sections of The Combustion Institute. The Combustion Institute, Funchal (Portugal), 6.7.1 - 6.7.4.

[13] Syri, S., 2002, *The impacts of climate change mitigation on air pollutant emissions*. In: Technology and climate change CLIMTECH 1999–2002; 14/2002 Final Report. Helsinki, Finland: The Finnish National Technology Agency, p. 224 - 30.

[14] Peña, F. and Blanco, J.M., 2007, Evaluation of the physical dew point in the economizer of a combined cycle burning natural gas, *Applied Thermal Engineering*, **27**, p. 2153 - 58.

[15] Barroso, J., Barreras, F. and Ballester, J., 2003, Behaviour of a High-capacity steam Boiler using Heavy Fuel Oil: Part I. High-temperature corrosion, *Fuel Process. Technol.* **86**, p. 89 - 105.

[16] Barreras, F. and Barroso, J., 2004, Behavior of a high-capacity steam boiler using heavy fuel oil Part II: Cold-end corrosion, *Fuel Process. Technol.* **86**, p. 107 - 21.

[17] Harris, B.W., 1990, Conversion of sulphur dioxide to sulphur trioxide in gas turbine exhaust. *Journal of Engineering for Gas Turbines and Power*, **112**, p. 585 - 89.

[18] Blanco, J.M., Mendía, F. and Peña, F., 2006, Comparative analysis of CO_2 and SO_2 emissions between combined and conventional cycles with natural gas and fuel oil consumption over the Spanish thermal power plants, *Fuel*, **85**, p. 1280 - 85.

[19] Bromly, J.H., Barnes, F.J. and Little, L.H., 1988, The effect of low levels of CO, H_2 and hydrocarbons on NO_2/NO ratios in heated gases. *Journal of the Institute of Energy,* **61**, 89-97.

[20] COMMISSION OF THE EUROPEAN COMMUNITIES. 1988. 88/609/EEC: Council Directive of 24 November 1988 on the limitation of emissions of certain pollutants into the air from large combustion plants. *Official Journal* no L 336, 07/12/1988 P. 0001.

[21] COMMISSION OF THE EUROPEAN COMMUNITIES. 1997. Proposal for a Council Directive relating to a reduction of the sulphur content of certain liquid fuels and amending Directive 93/12/EEC. *Office for Official Publications of the European Communities*, Luxembourg, 1-81.

[22] Chisholm, P.N. and Rochelle, G.T.,1999, *Ind Eng Chem Res* **38**, p. 4068 - 80.

[23] Cunningham, A.T.S., 1978, The reduction of atmospheric pollutants during the burning of heavy fuel oil in large boilers. *Journal of the Institute of Fuel*, **51**, p. 20 - 29.

[24] Miller, J.A., Branch, M.C., McLean, W.J. Chandler, D.W., Smooke, M.D., and Kee, R.J., 1985, In *20th Symp. (Int.) on Combustion*, The Combustion Institute, p. 673.

[25] Baukal, C.E., Heat transfer in industrial combustion, ISBN 0-8493-1699-5, CRC Press LLC, USA, 2000.

[26] Cotta, R.M., Mikhailov, M.D. and Özisik, M.N., 1986, Transient conjugated forced convection in ducts with periodically varying inlet temperature, *Int. J. Heat Mass Transfer* **30**, 10, p. 2073 - 82.

[27] Schmidt, F.W. and Willmot, A.J., *Thermal Energy Storage and Regeneration*, ISBN 0-07-055346-7, McGraw-Hill, New York, (1981).

[28] Park, P.M., Cho, H.C. and Shin, H.D., 2003, Unsteady thermal flow analysis in a heat regenerator with spherical particles, *Int. J. Energy Res.*, **27**, p. 161 - 72.

[29] Zarrinehkafsh, M.T. and Sadrameli, S.M., 2004, Simulation of fixed bed regenerative heat exchangers for flue gas heat recovery, *Applied Thermal Engineering*, **24**, p. 373 - 82.

[30] Demirbas, A., 1997, Calculation of higher heating values of biomass fuels, *Fuel* **76**, p. 431 - 34.

[31] The Ljungström Air Preheater 1920 An International Historic Mechanical Engineering Landmark Svenska Mekanisters Riksförening, *ASME*, 1920.

[32] Lipták, B.G., Process Measurements and Analysis, Instruments Engineers Handbook (Third edition), Chilton Book Co., Pennsylvania, USA, 1995.

[33] Aylward, G.H. and Findlay, T.J.V. SI chemical data (3rd ed.), Wiley, New York, 1994.

[34] Puigjaner, P. and Espuna, A.,1991, *Computer-oriented process engineering*, Elsevier, Amsterdam.

[35] Dekker, R. 1996, Applications of maintenance optimisation models: A review and analysis. *Reliab. Eng* **51**, p. 229 - 40.

[36] Robert, T. and Shahabudeen, P., 2004, Genetic algorithms for cost-effective maintenance of a reactor-regenerator system, *Int J. Adv. Manuf. Technol*, **23**, p. 846 - 56.

[37] Mills, A.F., *Heat Transfer* (second ed.), University of California, Los Angeles, California, 1999.

[38] Santos, W. F. and Trábelo, J.S., 1993, Transient conjugated forced convection in a parallel plate duct with convection from the ambient and periodic variation of inlet temperature, *Applied Scientific Research*, **51**, 3, p. 625 - 38.

[39] Li, W. and Sadik, K., 1991, Unsteady thermal entrance heat transfer in laminar with a periodic variation inlet temperature, *Int. J. Heat Mass Transfer* **34**, 10, p. 2581 - 92.

[40] Celata, G. P. and Carey, V. P., 2005, Heat Transfer and Transport Phenomena in Micro scale: *Proceedings of the International Conference on heat transfer*, p. 115 - 23.

[41] Zariffeh, E.K., Soliman, H.M. and Trupp, A.C., 1982, The combined effect of wall and fluid axial conduction on laminar heat transfer in circular tubes, *Heat Transfer*, **4**, p. 131 - 35.

[42] Variable sector plate quad sector air preheater, US Patent Issued on June 29, 1999.

[43] Flick, E., *Corrosion Inhibitors: An Industrial Guide*, William Andrew Inc, 1993.

[44] Skotheim, T. and Reynolds, R., *Conjugated Polymers: Processing and Applications,* CRC press, 2006.

[45] Schweitzer, P. E., *Paint And Coatings: Applications And Corrosion Resistance*, CRC press, 2006.

[46] Derek A. Bayliss, D.A. and Deacon, D., Steelwork Corrosion Control, Taylor & Francis, 2002.

[47] Bellan, J. and Elgobashi, S., 1985, Fuel composition effects on high-temperature corrosion in industrial/commercial boilers and furnaces: a review. *Journal of Engineering for Gas Turbines and Power*, **107**, p. 744 - 757.

[48] Harris, B.W., 1990, Conversion of sulphur dioxide to sulphur trioxide in gas turbine exhaust. *Journal of Engineering for Gas Turbine and Power*, **112**, 585-589.

[49] Uchida, S. and Tsuchiya, K., 1984, *Indus Eng Chem Process Design Develop* **23**, (2), p. 300 - 07.

[50] McCabe, W.L., Smith, J.C. and Harriott, P. *Unit operations of chemical engineering* (4th ed.), McGraw-Hill, New York, 1988.

[51] Weinell, C.E., Jensen, P.I., Dam-Johansen, K. and Livbjerg, H.,1992, *Ind Eng Chem Res* 31, p. 164 - 71.

[52] Armitage, J.W., Cullis, C.F., 1971, Studies of the reaction between nitrogen dioxide and sulphur dioxide. *Combustion and Flame*, **16**, p. 125 - 130.

[53] Bartok, W., Crawford, A.R., Skopp, A., 1971, Control of NO_X emissions from stationary sources. *Chemical Engineering Progress*, **67**, 2, p. 64 - 72.

[54] Demirbas, A., 1998, Determination of combustion heat of fuels by using non-calorimetric experimental data, *Energy Edu Sci Technol* **1**, p. 7 - 12.

[55] Demirbas, A., 2004. Combustion characteristics of different biomass fuels, *Progress Energy Combust. Sci.* **30**, p. 219 - 30.

[56] Chum, H.L. and Overend, R.P., 2001. Biomass and renewable fuels, *Fuel Process Technol* **71**, p. 187 - 95.

[57] Kuo, K.Y., *Principles of Combustion*. John Wiley and Sons, New York, 1986.

[58] Common Position (EC) No 52/2000 of 9 November 2000 adopted by the Council, acting in accordance with the procedure referred to in Article 251 of the Treaty establishing the European Community, with a view to adopting a Directive of the European Parliament and of the Council on the limitation of emissions of certain pollutants into the air from large combustion plants.

[59]http://europa.eu.int/comm/enlargement/negotiations/treaty_of_accession_2003/pdf/3_act_of_accession/aa00023_re03_en03.pdf

[60] Halkos, G., 1995, Evaluation of the Direct Cost of Sulphur Abatement Under the Main Desulphurization Technologies. *Energy Sources*, Vol. 17, p. 391-412.

[61] E. M. Goodger, *Transport fuels technology*, Landfall press, (2006).

[62] Blanco, J.M. and Peña, F., 2006. Obtención del valor real de las pérdidas de difícil evaluación, aplicables al cálculo del rendimiento de calderas, *Información tecnológica*, Vol. 17, 3, p. 123 - 28.

[63] Bejan, A., Tsatsaronis, G. and Moran, M., *Thermal design and optimization*, John Wiley & Sons, Inc, New York, 1996.

In: Acid Rain Research Focus
Editor: Lyman G. Roglesfield, pp. 45-73
ISBN 978-1-60456-373-3

Chapter 2

WET REMOVAL OF ATMOSPHERIC PARTICLES BY RAIN

Constantin Andronache
Boston College, Chestnut Hill, Massachusetts, USA

Abstract

Acid precipitation represents the deposition of acidic components in rain, snow, dew, or dry particles. Acid rain occurs when sulfur dioxide and nitrogen oxides are emitted into the atmosphere, undergo chemical transformations, and are absorbed by water droplets in clouds. The droplets fall to Earth and increase the acidity of the soil, lakes and streams. Acid rain can have adverse impacts on forests, freshwaters and soils, insects, aquatic life forms, and can cause damage to buildings. In absence of precipitation, acid deposition occurs via dry deposition, a significant process, especially in dryer climates, or during periods of drought. Wet deposition of acids occurs when precipitation removes acids from the atmosphere and delivers it to the Earth's surface. This can result from the deposition of acids produced in the raindrops or by the precipitation removing the aerosol particles either in-cloud or below-cloud. In this chapter, we address in detail the wet removal of aerosol particles by rain. Wet removal of atmospheric aerosols by falling precipitation is a major removal mechanism for airborne particles. In-cloud wet removal is the dominant mechanism and the rate of aerosol removal in cloud is largely dependent on rainfall rate and the collision efficiency between raindrops and aerosol particles incorporated in cloud droplets. Below-cloud removal rate of aerosols by rainfall is dependent on the rainfall rate, collision efficiency between raindrops and aerosol particles, and on aerosol size distribution. Results presented show the link between the aerosol removal rate and the microphysical characteristics of precipitation and aerosol such as raindrop size distribution, aerosol size spectrum, collision mechanisms, and environmental variables.

1. Introduction

Atmospheric particles from man-made and natural sources exhibit a wide range of interactions with gaseous chemical species, electromagnetic radiation, clouds and precipitation. Aerosols impact visibility, air quality, cloud formation, and are considered a key factor in the climate system (Ramanathan et al., 2001; Adams and Seinfeld, 2002; Lohmann and

Feichter, 2005; Ghan and Schwartz, 2007). Aerosols originate mainly from Earth's surface sources as primary particles (emissions from soil, biosphere, sea, combustion, and volcanic eruptions). A smaller source of primary particles is from aircraft engines, and cosmic origin. Aerosols are also created in atmosphere as a result of gas-to-particle conversion (secondary particles). Atmospheric aerosols are transported by the air flow and are involved in various gas - aerosol, aerosol - aerosol, and aerosol - hydrometeor interactions before eventually they are removed from atmosphere by dry and wet deposition. Dry deposition is the process by which atmospheric trace chemicals are transferred by air motions to the surface of the Earth. Gravitational settling affects deposition of particles, especially those larger than a few micrometers in diameter (Sehmel, 1980). Wet deposition is caused by removal of particles from atmosphere during precipitation. The acidity of precipitation is mainly produced by the significant anthropogenic emissions of SO_2 and NO_x leading to sulphate and nitrate particles. These aerosols, are generated in atmosphere by gas-to-particle conversion processes. It is estimated that wet deposition accounts for $\sim$ 60 % of the total deposition of sulfur. Emissions of sulphur and nitrogen oxides contribute 70 % and 30 % respectively to the acidity of precipitation (Jennings, 1998). Acid rain formation involves a series of processes: SO_2 and NO_x emissions, transport by atmospheric circulation, gas phase chemistry, interactions with aerosols and hydrometeors, and removal by dry and wet deposition. Among these processes, the wet removal of sulphates and nitrates by precipitation is of great practical importance and requires long term monitoring (Hicks, 2005) and numerical simulations using local, regional and global air quality models (Gong et al., 2003; Grell et al., 2005; Mueller et al., 2006; Tost et al., 2006; Debry et al., 2007; Tombette and Sportisse, 2007). In this effort, experimental measurements and model estimations of the removal rates of aerosols represent an important step.

Generally, the term aerosol scavenging by precipitation denotes the removal of pollutants from the air by either rain or snow. Rainout (or snowout), which is the in-cloud capture of particulates as condensation nuclei, is one form of scavenging. The other form is washout, the below-cloud capture of particulates and gaseous pollutants by falling raindrops. Large particles are most efficiently removed by washout. Small particles (especially those less than 1 μm in diameter) more easily follow the air stream flowing around raindrops and generally avoid capture by raindrops except in heavy rain events. The scavenging of aerosol particles by dry and wet deposition represent important processes by which a balance is maintained between the sources and sinks of airborne particles. The wet removal by precipitation is the most efficient atmospheric aerosol sink and the detailed mechanism of this process involves microphysical interactions between aerosols and hydrometeors (Pruppacher and Klett, 1998; Seinfeld and Pandis, 1998; Jacobson, 2003). Aerosol wet removal is typically represented in current models by scavenging coefficients in aerosol mass continuity equations. These scavenging coefficients are often expressed as a function of bulk quantities available from measurements, such as the precipitation rate.

Significant experimental and theoretical work has been done to estimate the aerosol scavenging coefficients under a variety of conditions (Dana and Hales, 1976; Wang and Pruppacher, 1977; Grover et al., 1977; Wang et al., 1978; Radke et al., 1980; Scott, 1982; Slinn, 1983; Ten Brink et al., 1987; Jylhä, 1991; McGann and Jennings, 1991; Sparmacher et al., 1993; Garcia Nieto et al., 1994; Okita et al., 1996; Mircea et al., 2000; Jung et al., 2003; Andronache, 2003; Chate and Pranesha, 2004; Laakso et al., 2003; Zhang et al., 2004;

Latha et al, 2005; Maria and Russell, 2005; Bae et al., 2006; Tost et al., 2006; Zhao and Zheng, 2006; Feng, 2007). Recent reports show that the representation of aerosol removal processes in current aerosol transport models remains a source of uncertainty (Rasch et al., 2000; Tost et al., 2006; Sportisse, 2007). This is due to multiple scales involved in aerosol-cloud interactions and the limited ability of current models to predict the precipitation field. Because the understanding of wet removal processes remains crucial in local and regional pollution studies, efforts continue in this area of research, combining experimental work, and model validation with increasing resolution data from field measurements.

An aerosol particles can be removed from atmosphere by precipitation as a result of two main processes. The first process involves nucleation scavenging of the aerosol particles which serve as cloud condensation nuclei (CCN) or ice nuclei (IN) in the initial stage of cloud formation. As a result of the nucleation scavenging, some aerosol particles become cloud droplets and can be removed from atmosphere after in-cloud collection by falling raindrops. This process is often called in-cloud scavenging and depends strongly on rainfall intensity as we will describe below. The nucleation scavenging processes are controlled by the ambient water vapor supersaturation, aerosol size and chemical composition. The second process is represented by the collection of aerosol particles present in the boundary layer (BL) by falling raindrops. This process depends on the net action of various forces influencing the relative motion of aerosol particles and hydrometeors. It is strongly dependent on aerosol size, and rainfall intensity.

Nucleation scavenging dominates the removal processes at the beginning of cloud formation and experimental and model results show that this process can deplete the original aerosol population in air up to 75% – 90 % (Flossmann et al., 1985, 1987; Ten Brink et al., 1987; Schumann, 1991). While a large fraction of BL aerosol is incorporated in-cloud by nucleation scavenging, not all of these particles are removed in-cloud by falling raindrops. Some cloud droplets evaporate, releasing aerosol back into environment. Also some falling raindrops evaporate before reaching the ground (especially those with diameter less than 0.1 mm). Generally, aerosols are cycled in clouds several times before reaching ground. Nevertheless, the efficiency of nucleation scavenging, combined with intense or persistent precipitation, leads to significant aerosol depletion from the atmosphere.

Similar mechanisms act during in-cloud scavenging of cloud droplets, and during below-cloud scavenging of aerosol. These mechanisms are: convective Brownian diffusion, interception, inertial impaction, thermophoresis, diffusionphoresis, airflow turbulence and electrostatic interaction. A literature review on these processes is given by Pruppacher and Klett (1998). One significant difference between the in-cloud and below-cloud scavenging processes is that the in-cloud activated aerosol grows typically into cloud droplet (with a diameter of about 10 μm), which makes the collection by falling raindrops very efficient (and largely dominated by inertial impaction). In contrast, aerosol particles below-cloud have much smaller sizes and their collection by falling raindrops is less efficient (small aerosol scavenging is dominated by Brownian diffusion, while the scavenging of coarse aerosols is dominated by inertial impaction). We note however that while the below-cloud removal process is less efficient than the in-cloud removal process, there are several environmental conditions when the below-cloud removal process is important, especially in cases when mechanical generated large particles are present near the Earth's surface (Andronache, 2003; Feng, 2007). In this respect both in-cloud and below-cloud scavenging processes must be represented in air quality and climate models (Rasch et al., 2000; Henzing et al., 2006; Tost et al., 2006).

Given the complexity of aerosol scavenging processes, there is no simple general theory to account for all aerosol sizes and for all atmospheric conditions. Particularly, the variability of phoretic forces, electric charge and turbulence tend to affect the below-cloud scavenging of airborne particles and this variability is reflected in the reported experimental and model estimated values of the below-cloud scavenging coefficient. The purpose of this chapter is to present the general framework for the estimation of scavenging of aerosol relevant for air quality, and climate models with particular applications to acid rain problem. The practical outcomes of this review are the scavenging coefficients that can be used in numerical models to describe the atmospheric aerosol changes due to rain. The presented results are suitable for two modeling approaches. One approach is the study of the evolution of detailed bin resolved aerosol mass size distribution (as in the case of analysis of high resolution aerosol measurements from field experiments). The second approach is the study of evolution of the total aerosol mass concentration during rain events (as in the case of local and regional models used in pollution and acid rain studies). First, we describe the scavenging rate concept, and the required parameters for its calculation. Then we show results for below-cloud and in-cloud scavenging rates determined from measurements and from modeling studies. We conclude with a summary of the main charcteristics of the scavenging rate and with remarks on areas that need further research.

2. Scavenging Rate

The method used to evaluate the scavenging rate of aerosols, or the scavenging coefficients, is based on the concept of efficiency of collision between an aerosol particle and a falling raindrop (Slinn, 1983; Pruppacher and Klett, 1998; Seinfeld and Pandis, 1998). When a raindrop of diameter D falls below-cloud, it creates an airflow with streamlines that deviates from the drop surface. Some of the aerosol particles present in the swept out volume of the raindrop collide with the raindrop and are collected. The rate of mass concentration change of all particles of diameter d is

$$\frac{dn_M(d)}{dt} = -\lambda(d)n_M(d) \tag{1}$$

where: $n_M(d)$ is the mass size distribution of aerosol, and the scavenging rate λ has expression

$$\lambda(d) = \int_0^\infty \frac{\pi}{4} D^2 U_t(D)E(d,D)N(D)dD \tag{2}$$

where, $U_t(D)$ is the raindrop terminal velocity, $N(D)$ is the raindrop size distribution, and $E(d,D)$ is the collection efficiency. The raindrop terminal velocity is taken as a fit of experimental data of Gunn and Kinzer (1949), as shown below. The raindrop size distribution is represented by an empirical Marshall and Palmer (1948) exponential fit, or by a gamma function as described in the next section. The method outlined here can be used to estimate below and in-cloud scavenging coefficients as will be illustrated in section 3. The parameters involved in scavenging rate estimation change in time and with altitude, and cause significant variability in λ. Our main effort is to describe scavenging by liquid precipitation. In the next subsections, we review the main characteristics of $N(D)$, $U_t(D)$, and $E(d, D)$ needed to estimate the scavenging rate.

2.1. Raindrop Size Distribution

The raindrop size distribution (DSD) describes the number of raindrops with diameter between D and $D + dD$. DSD varies in time and space, and depends on precipitation type. Typical raindrops are oblate spheroids, and in the description of DSD it is assumed that D is the equivalent diameter, or the diameter of a sphere with the volume equal with that of the deformed drop. Generally, three quantities are important for the characterization of rain: (1) the size distribution $N(D)$, the number of raindrops per cubic meter of air per mm size interval; (2) the water content, w_L given by

$$w_L(gm^{-3}) = (\frac{\pi}{6}) \times 10^{-3}\rho_w \int_0^\infty N(D)D^3dD \tag{3}$$

with D in mm and ρ_w in g cm^{-3}; and (3) the rainfall rate, R expressed in mm h^{-1}

$$R = 6\pi \times 10^{-4} \int_0^\infty N(D)U_t(D)D^3dD \tag{4}$$

where, U_t is the terminal velocity of raindrop, in ms^{-1}.

Marshall and Palmer (1948) (MP) analyzed the observations of Laws and Parsons (1943), and introduced an exponential fit of DSD

$$N(D) = N_0 \times exp(-\Lambda D) \tag{5}$$

where $N(D)$ is in m^{-3} mm^{-1}, D is in mm, rainfall rate R is in mm h^{-1}, N_0 = 8000 m^{-3} mm^{-1}, and Λ = 4.1 $R^{-0.21}$ mm^{-1}. Generally, the Marshall - Palmer DSD overestimates the number of very small raindrops as well as the number of large raindrops. Studies by Joss et al. (1968) showed that a MP exponential fit of data varied with precipitation type. For the MP raindrop size distribution, the above rainfall intensity equation becomes

$$R = N_0 6\pi \times 10^{-4} \int_0^\infty exp(-\Lambda D)U_t(D)D^3dD \tag{6}$$

Thus, for a given rain rate, N_0 and Λ cannot be independently specified. This shows that when MP type distributions are fitted to experimental data, rather than fitting N_0 and Λ independently, the fit should be constrained to allow the rainfall rate to be self-consistent (Bennett et al., 1984).

A series of more detailed studies showed that N_0 can not be considered constant, but rather a function of R (Joss et al., 1968; Waldvogel, 1974). While MP type DSD is attractive due to its simplicity, it fails to represent observed instantaneous spectra (for sampling time less 1 min). DSD tends to be exponential only when the sampling time is sufficiently long. Figure 1a shows the MP fit from observations, contrasting widespread, drizzle and thunderstorm precipitation. Figure 1b shows that the increasing rainfall intensity R, leads to change in DSD slope, while the intercept parameter is assumed unchanged. In fact, observations show that N_0 changes with rain type and rain intensity R (Joss et al., 1968; Waldvogel,

Table 1. Examples of observed variation in N_0 and Λ during different types of rain.

N_0 (m^{-3} mm^{-1})	Λ (mm^{-1})	Rain type	Notes
8000	$4.1R^{-0.21}$	Widespread	(a)
30000	$5.7R^{-0.21}$	Drizzle	(b)
7000	$4.1R^{-0.21}$	Widespread	(c)
1400	$3.0R^{-0.21}$	Thunderstorm	(d)
$7000R^{0.37}$	$3.8R^{-0.14}$	Thunderstorm	(e)

(a) Marshall and Palmer (1948; (b)-(d) Joss et al. (1968); (e) Sekhon and Srivastva (1971).

1974; Sekhon and Srivastva, 1971). In this respect, MP DSD is useful in scavenging calculations representative for widespread precipitation events or for average conditions over the area of interest. For the characterization of detailed scavenging process when R changes rapidly in time and space, it is necessary to consider a DSD function which can describe such variability.

Other DSD have been proposed, and the three-parameter gamma distribution is one of the most preferred when fitting observations, especially since the rapid advances in rain measurements by radar (Ulbrich, 1983; Ulbrich and Atlas, 1998). This DSD is represented as

$$N(D) = N_{0g} D^{\mu_g} exp(-\Lambda_g D) \tag{7}$$

where $N(D)$ is in m^{-3} mm^{-1}, D is in mm, N_{0g} m^{-3} $mm^{-1-\mu}$, and Λ_g is in mm^{-1}. Other mathematical forms of DSD have been employed, but the advantages of the distribution from above equation are that with three parameters it is general enough to describe the fluctuations in the DSD observed on small space and timescale, and includes the exponential distribution as a special case. For rain DSD retrieval, the problem is how to determine the three parameters (N_{0g}, μ_g, Λ_g) from radar measurements. Zhang et al. (2001) showed that the three parameters can be determined from radar measurements of: reflectivity (Z_{HH}), differential reflectivity (Z_{DR}), and a constrained relation between shape (μ_g) and slope (Λ_g) derived from video disdrometer observations.

These authors found

$$\mu_g = -0.016\Lambda_g^2 + 1.213\Lambda_g - 1.957 \tag{8}$$

where Λ_g is in mm^{-1}. They determined also a relation between R, N_{0g}, Λ_g and μ_g in the following form

$$R = 7.125 \times 10^{-3} N_{0g} \Lambda_g^{-(3.67+\mu_g+1)} \Gamma(3.67 + \mu_g + 1) \tag{9}$$

where R is in mm h^{-1}, N_{0g} is in m^{-3} $mm^{-\mu-1}$, and Λ_g is in mm^{-1}. Figure 2a shows the gamma DSD for a selection of parameters (Λ_g, μ_g) constrained by equations (8) and (9), for $R = 1$ mm h^{-1}. We note that the shape and slope of DSD can vary dramatically. Figure 2b illustrates variations of DSD with rain intensity R. This plot confirms the N_{0g} varies with R as indicated by equation (9) as observed earlier by Joss et al. (1968). We must note that parameters of the gamma DSD have also significant variability, and the parameters used here are not universal. Recent advances in radar technology and its wide use in precipitation measurements and air pollution studies, make the gamma type DSD an appropriate choice for aerosol scavenging studies.

2.2. Raindrop Terminal Velocity

In warm clouds, raindrops form by growth of cloud droplets. Small cloud droplets grow by condensation when the environment is supersaturated with respect to liquid water. Condensation itself is not sufficient to produce raindrops in a time frame observed in precipitating clouds. After the initial important diffusional growth of small cloud droplets, the collision and coalescence of cloud droplets lead to raindrop formation. In freezing clouds, raindrops can also form by melting of ice particles falling from higher altitudes in the cloud. Raindrops fall due to gravity, and the drag force is acting to reduce the speed of fall and eventually the raindrops reach a terminal velocity $U_t(D)$. Raindrop velocity before reaching the terminal velocity can be also important in experimental work where measurements are conducted in laboratory using installations that do not permit the simulated raindrops to reach terminal velocity. Here we review some relevant findings about terminal velocity dependence on raindrop diameter D, and its variation with altitude.

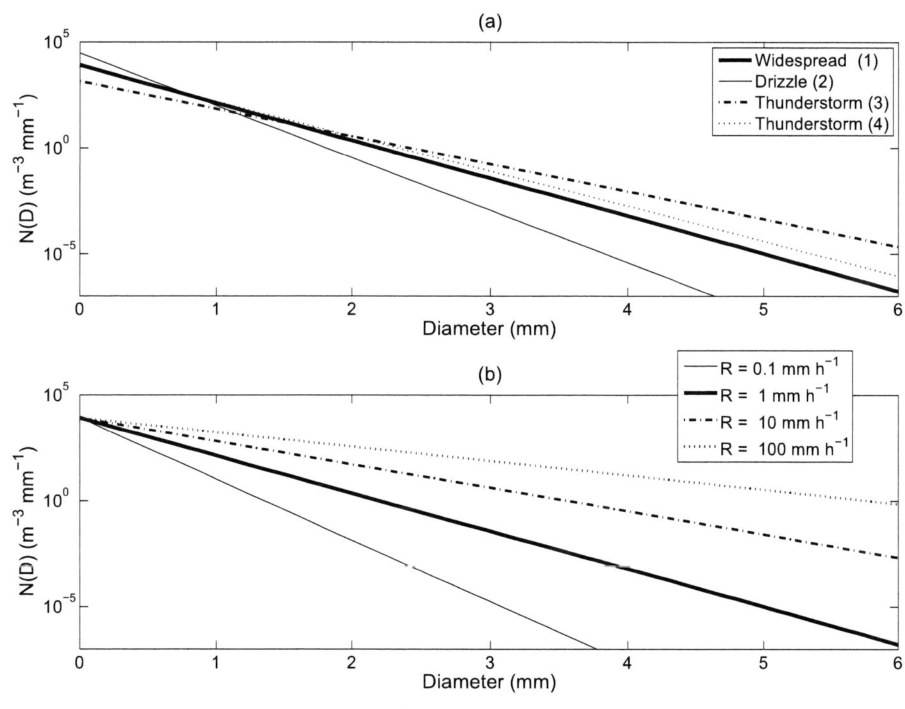

Figure 1. Exponential raindrop size distribution: (a) for various rain types (1) - Marshall and Palmer, 1948; (2) and (3) - Joss et al., 1968, (4) - Sekhon and Srivastva, 1971); (b) widespread precipitation, for various rainfall intensity.

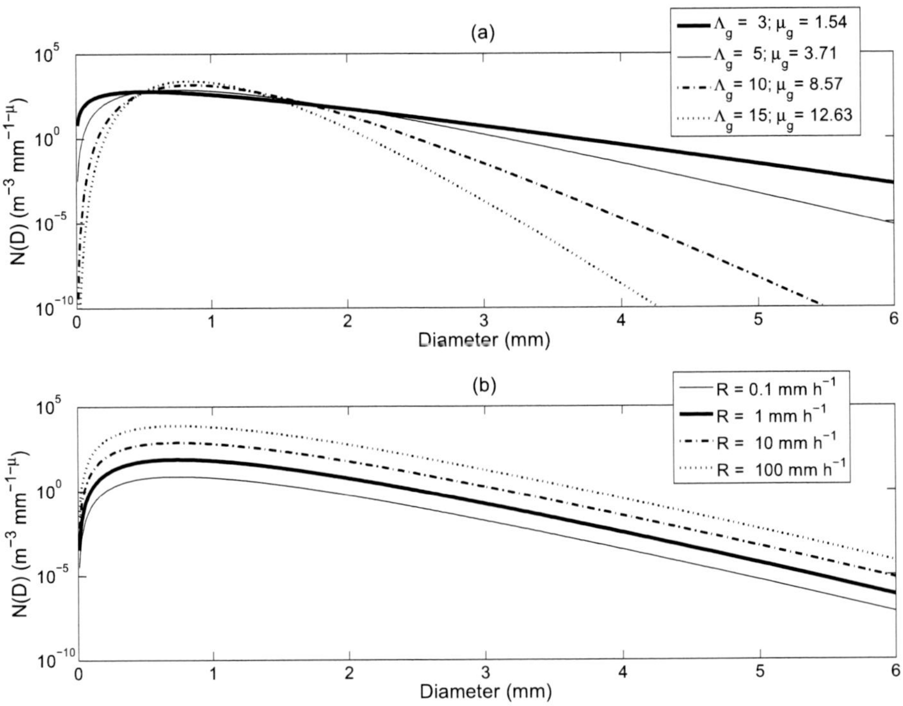

Figure 2. Gamma raindrop size distribution: (a) for $R = 10$ mm h^{-1} and several parameters of Λ_g and μ_g, and (b) for $\Lambda_g = 5$, $\mu_g = 3.71$, for several rainfall rates, R.

The terminal velocity of raindrops has been determined experimentally and theoretically. Experimental determination of the raindrop terminal velocity was reported by Gunn and Kinzer (1949), and their data was fitted with various functions used in scavenging studies (Table 2). Figure 3 shows some of the best fitting curves of the Gunn and Kinzer data reported in literature. Other approximations have been used in cloud physics and applied in aerosol scavenging studies. For instance, Kessler (1969) proposed $U_t(D) = 130 \times D^{0.5}$ (where U_t is in ms^{-1} and D is in m). Similarly, Atlas and Ulbrich (1977) used $U_t(D)$=cD^{γ}, with $c = 3.778$ and γ=0.67 (with U_t in ms^{-1} and D in mm). These approximations, are attractive for their simple form, for the fact that $U_t = 0$ when $D = 0$, and for use outside the range of D available from the Gunn and Kinzer measurements (in their experiment D is varied between 0.1 mm and 5.8 mm).

Theoretical computations of the terminal velocity are based on solving the problem of raindrop equilibrium in air at various Reynolds numbers. The raindrop falls under the forces of gravity and drag. These forces change the shape of the raindrop, leading to shape distortion which in turn, changes the drag force. For small raindrops ($D < 0.05$ mm), the surface tension is strong enough to keep the drop shape close to spherical. The flow around the drop is considered to be laminar, and in this case, the Stokes law applies: the acceleration of the raindrop is determined by equation

Table 2. Selected expressions of the raindrop terminal velocity $U_t(D)$ (in ms^{-1} units) as a function of raindrop diameter D (in mm units).

Reference	Formula
Best (1950)	$U_t(D) = 9.58[1 - exp(-(D/1.77)^{1.147})]$
Foote and Du Toit (1969)[a]	$U_t(D) = \sum_{j=0}^{N} a_j D^j$
Dingle and Lee (1972)[b]	$U_t(D) = \sum_{j=0}^{M} b_j D^j$
Uplinger (1977)	$U_t(D) = 4.85 D exp(-0.195D)$
Gossard et al. (1992)	$U_t(D) = 9.65[1 - exp(-0.53D)]$

[a]Foote and Du Toit (1969) provided a polynomial fit of the Gunn and Kinzer (1949) experimental data, using a polynomial of degree $N = 9$, where $a_0 = -8.5731540 \times 10^{-2}$; $a_1 = 3.3265862$; $a_2 = 4.3843578$; $a_3 = -6.8813414$; $a_4 = 4.7570205$; $a_5 = -1.9046601$; $a_6 = 4.6339978 \times 10^{-1}$; $a_7 = -6.7607898 \times 10^{-2}$; $a_8 = 5.4455480 \times 10^{-3}$; $a_9 = -1.8631087 \times 10^{-4}$;
[b]Dingle and Lee (1972) provided a fit of the same data using a polynomial of degree $M = 3$, where $b_0 = -0.166033$; $b_1 = 4.91844$; $b_2 = -0.888016$; $b_3 = 0.054888$;

$$m\frac{dU}{dt} = F_g - F_d \tag{10}$$

where, $F_g = g\rho_w \pi D^3/6$ is the force of gravity, $F_d = 3\pi D\mu_a U$ is the friction force, g is the gravity acceleration, ρ_w is raindrop water density, μ_a is the dynamic viscosity of air, and m is the raindrop mass. At equilibrium, the terminal velocity is

$$U_t = \frac{\rho_w g D^2}{18\mu_a} \tag{11}$$

For drops larger than about 0.1 mm, the flow around the raindrop becomes turbulent and Stokes law does not apply. The transition to a turbulent flow regime is characterized by the Reynolds number $R_e(D) = \frac{\rho_a U(D) D}{\mu_a}$, which expresses the ratio of turbulent to laminar forces. For large raindrops, the aerodynamic differences around the drop will distort the drop shape, such that the vertical dimension will decrease and the horizontal dimension will increase (producing an oblate spheroid) (McDonald, 1954; Beard et al., 1991). For drops larger than 2 mm, the terminal velocity increases with diameter much less than it would be without shape distortion. Following van Boxel (1997), the acceleration of raindrop of any diameter D can be determined from equation

$$m\frac{dU}{dt} = F_g - F_d \times G(R_e, W_e) \tag{12}$$

where G is a function of Reynolds number R_e and Weber number W_e. The Weber number is defined as $W_e = \frac{\rho_a U(D)^2 D}{\sigma}$, where σ is the raindrop surface tension. The model was fitted to agree with the terminal velocities measured by Gunn and Kinzer (1949). Based on this model, Figure 4a shows that terminal velocity is reached in less than a second by raindrops with $D < 1$ mm, while raindrops with diameter $D = 3$ mm or $D = 5$ mm requires longer time to reach terminal velocity. The distance traveled by falling raindrop is shown in figure 4b. The larger raindrops need about 12 m to reach the terminal velocity in atmosphere at sea level conditions. Very small raindrops ($D < 0.5$ mm or so) can reach terminal velocity in less than 2 m. Thus for laboratory studies of aerosol scavenging, where the terminal velocity is not reached, it is important to estimate the raindrop velocity based on a model as the one described by van Boxel (1997).

Another important aspect of the raindrop terminal velocity, necessary in scavenging studies, is its variation with altitude. Beard (1985) determined an expression for $U_t(D)$ as a function of atmospheric density and temperature

$$U_t(z) = U_t(0) \times \left(\frac{\rho_a(0)}{\rho_a(z)}\right)^{m(D)} \tag{13}$$

where, $\rho_a(0)$ is air density at sea level, $\rho_a(z)$ is air density at altitude z, and $m(D) = 0.375 + 0.025D$, with D in mm. Figure 5 illustrates the increase of the terminal velocity of raindrops with altitude, shown as atmospheric pressure level. For larger raindrops, the terminal velocity increases significantly with altitude.

2.3. Collection Efficiency

A crucial parameter in characterization of the scavenging rate is the collection efficiency between falling raindrop and aerosol particles. The general formulation of the collection efficiency allows its application both in-cloud and below-cloud scavenging processes, with special attention to the dominant factors. Thus, we focus here on the work done during the last decades to characterize the below-cloud scavenging based on the concept of efficiency of collision between a falling raindrop and aerosol particles. Later, when we describe the in-cloud scavenging we will note the specific differences. Earlier work by Greenfield (1957) recognized the role of Brownian motion in increasing efficiency of collision between the raindrop and the very small aerosol particles (with diameters $d < 0.01\ \mu m$). Similarly, large particles (with diameters $d > 2\ \mu m$) tend to have high collision efficiency because of their inertia.

The particles with diameters in the range $[0.01 - 2]\ \mu m$ appeared to have a very small efficiency of collision and this size range is named *scavenging gap*. Further work by Slinn and Hales (1971) showed that thermophoresis could enhance the below-cloud scavenging of aerosols with diameters in the range $[0.01 \ - 1]\ \mu m$. Experimental work and model results by Grover et al. (1977), Wang et al. (1978), Byrne and Jennings (1993), showed that the presence of charge on aerosol particles and raindrops can substantially increase the below-cloud scavenging of aerosols with sizes in the *scavenging gap*. More recent model estimations by Tinsley et al. (2000) indicate that below-cloud scavenging can be increased significantly for aerosols resulted from evaporation of charged cloud drops. Additional evidence shows that aerosol-cloud interactions are significantly affected by electric charge in a wide range of processes (Pruppacher and Klett, 1998; Tripathi and Harrison, 2002; Carslaw et al., 2002; Harrison and Carslaw, 2003).

$E(d, D)$ is the collection efficiency, which is assumed to be equal with the collision efficiency (Slinn, 1983). The experimentally accessible quantity is the "collection efficiency", which is the ratio of the actual cross-section for aerosol coalescence to the geometric cross section. Various models treating aerosol scavenging calculate the "collision efficiency". The collision of two drops of comparable size is not always followed by coalescence, and the two drops might not stick together. The coalescence depends on particle shape, composition, and ambient relative humidity. However, it has been shown (Pruppacher and Klett, 1998) that a collision is followed by coalescence for $d/D \ll 1$, and in this case, the "collection efficiency" is equal with the "collision efficiency". This is because the kinetic energy of tiny aerosol particles is relatively small in comparison with that of large particles (falling raindrops), and the probability that bounce-off occurs when small aerosol particles collide with raindrops is low. In this sense, the coalescence efficiency of aerosol colliding with raindrops is approaching unity. This condition is satisfied by typical aerosol present in the BL below-cloud, where aerosol particles uptake water, become wet, and play the role of small spherical particles that collides with a falling raindrop. Exceptions might be found in cases of predominant hydrophobic aerosols that have cluster shapes, and do not became wet and spherical during precipitation. For these cases, this theory does not apply.

The collection efficiency is given by $E = E_b + E_{int} + E_{imp}$ (Slinn, 1977, 1983; also summarized by Seinfeld and Pandis, 1998), where

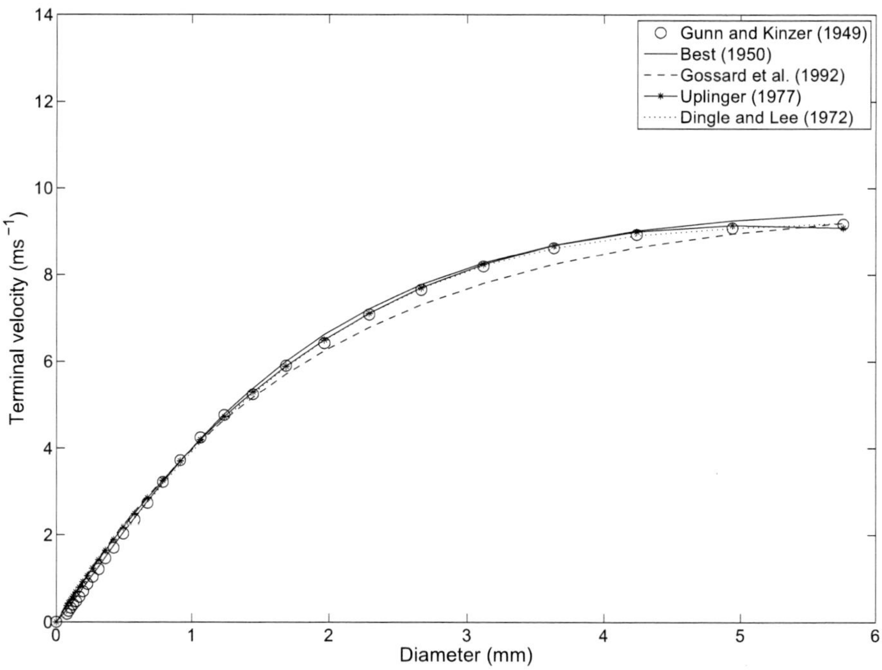

Figure 3. Terminal velocity of raindrop. The experimental observations of Gunn and Kinzer (1949) data are shown in open circles. Curves represent several proposed fitting expressions of the observations.

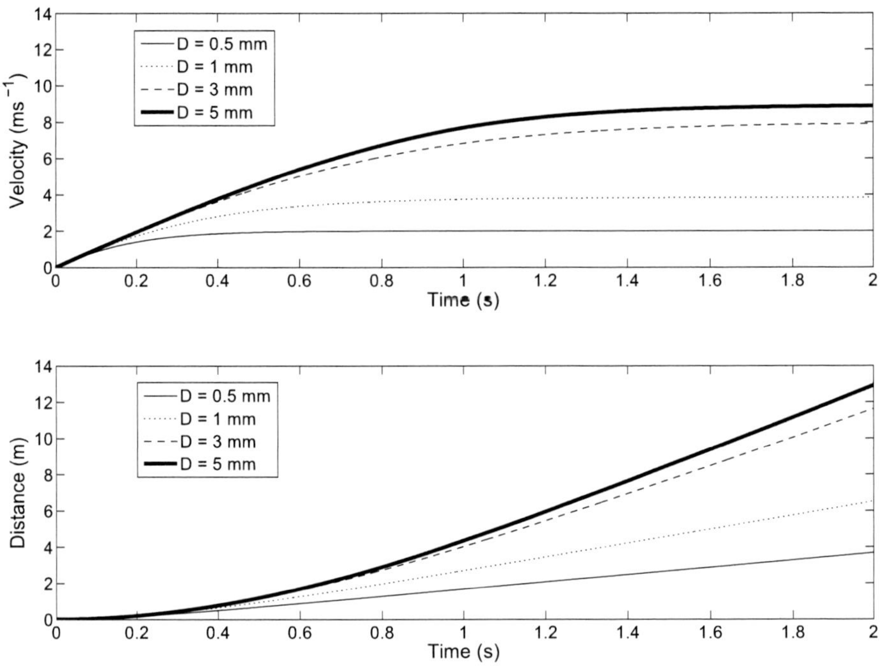

Figure 4. (a) Raindrop velocity versus time, and (b) distance versus time, for several raindrop diameters.

$$E_b(d, D) = \frac{4}{R_e S_c}(1 + 0.4R_e^{1/2}S_c^{1/3} + 0.16R_e^{1/2}S_c^{1/2}) \tag{14}$$

is the Brownian diffusion contribution, especially important for particles with $d < 0.01\ \mu m$

$$E_{int}(d, D) = 4\frac{d}{D}\left[\frac{\mu_a}{\mu_w} + (1 + 2R_e^{1/2})\frac{d}{D}\right] \tag{15}$$

is the contribution of interception term, and

$$E_{imp}(d, D) = \left[\frac{S_t - S^*}{S_t - S^* + 2/3}\right]^{3/2}\left(\frac{\rho_p}{\rho_w}\right) \tag{16}$$

is the contribution of inertial impaction to the total collection efficiency. In this last equation $S^* = \frac{1.2+1/12ln(1+R_e)}{1+ln(1+R_e)}$. Thermophoretic, diffusionphoretic and electric contributions were presented and discussed in literature (Davenport and Peters, 1978; Pruppacher and Klett, 1998; Andronache 2004a; Chate and Devara, 2005). Other notations in the above equations are: R_e is the raindrop Reynolds number, $S_c = \frac{\mu_a}{\rho_a D_a}$ is the aerosol Schmidt number, ρ_a and μ_a are the air density and viscosity, $D_a = \frac{k_B T_a C_c}{3\pi\mu_a d}$ is aerosol diffusivity in air, k_B is the

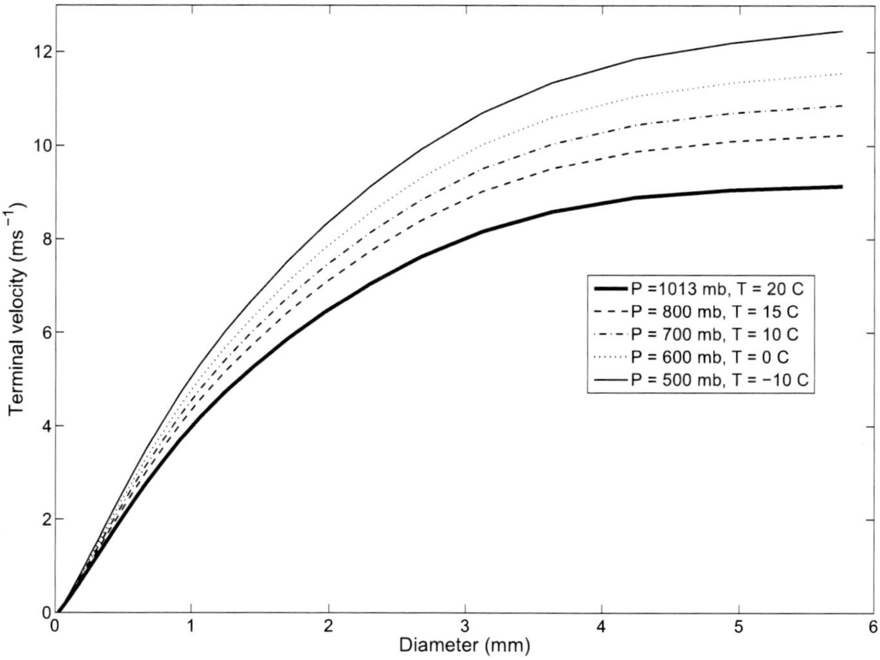

Figure 5. Terminal velocity of raindrop for various atmospheric pressure levels.

Boltzmann constant, T_a is the air absolute temperature, ρ_p is the aerosol particle density, C_c is the Cunningham slip correction factor to account for non-continuum effects associated with small particles, $C_c{=}1{+}\frac{2\lambda_a}{d}[1.257{+}0.4\exp(-\frac{1.1d}{2\lambda_a})]$, and λ_a is the mean free path of air molecules.

Figure 6 illustrates the role of various terms that contribute to the collision efficiency, E, between a raindrop of diameter D = 0.1 mm and an aerosol particle of diameter d: the Brownian diffusion dominates E for particles with diameters $d < 0.01$ μm, while the interception and impaction become important for larger particles with $d > 1$ μm. The effect of raindrop size on the collision efficiency is illustrated in Figure 7. Thus, small aerosol particles are collected more efficient by smaller raindrops than by large raindrops. For coarse aerosol particles, due to important inertial term, there are fewer differences in E for various raindrop sizes.

It has been shown that E is significantly enhanced for aerosol particles with diameters d in the range $[0.1 - 2]$ μm by phoretic forces, and turbulence, not included above (Slinn and Hales, 1971; Grover et al., 1977; Wang et al., 1978; Byrne and Jennings; 1993; Tinsley et al., 2000). The effects of phoretic and electric forces are not well known in part because

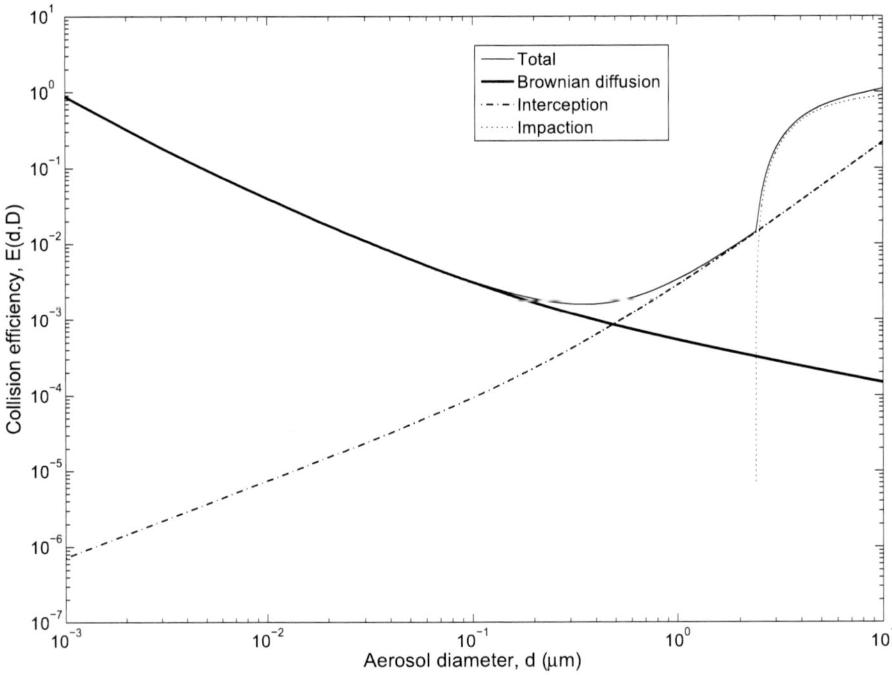

Figure 6. Various contributions to the efficiency of collision between a raindrop of diameter $D = 0.1$ mm and an aerosol particle of diameter d.

of lack of measurements of electric charge, and the temperature difference between falling raindrop surface and ambient air. Further work is required to quantify these effects.

3. Model Estimations and Observations

3.1. Below-Cloud Scavenging

Many studies have been performed to measure and estimate the below-cloud scavenging coefficient. This is mainly justified by the following: a) There is significant interest in understanding how small aerosol particles (with diameter less than 1 μm) emitted from Earth's surface or from the BL, by pollution sources for example, are removed by precipitation (Loosmore and Cederwall, 2004; Sportisse, 2007); b) In laboratory studies it is possible to control the falling water drop diameter D, the aerosol particle size d, and the ambient conditions such as air temperature, pressure, and relative humidity. Using such detailed controlled experiments, it was possible to obtain experimental data and correlations between E and relevant parameters that can be used in atmospheric models (Pruppacher and Klett, 1998); c) In many experiments, it was possible to control the aerosol properties (such as diameter and physical properties), during real precipitation events. In such cases, particles are released in controlled manner and the effect of precipitation is investigated (Nicholson et al., 1991; Sparmacher et al., 1993); d) It is somehow practical and more affordable to study

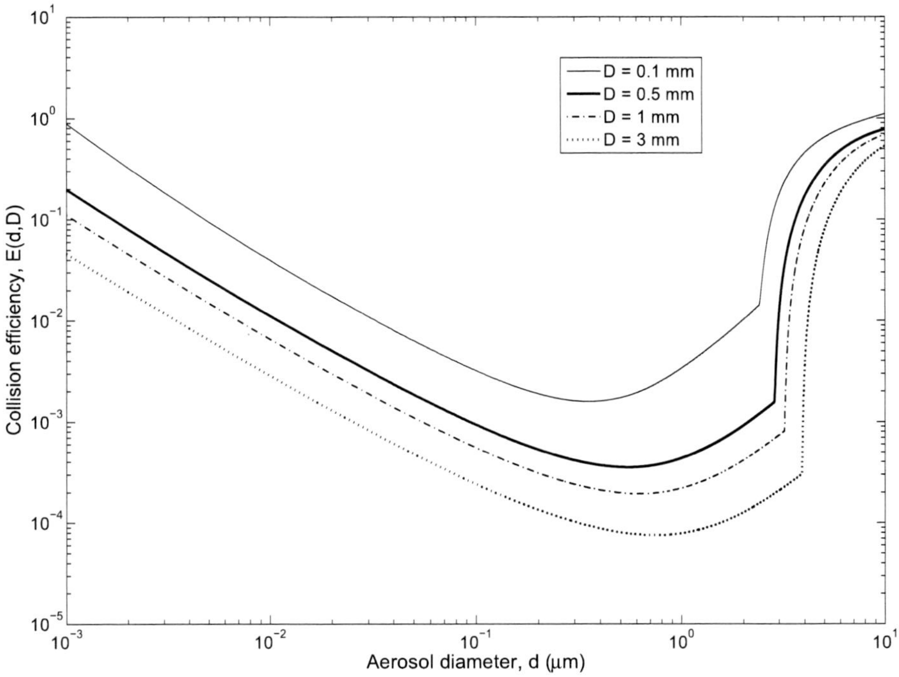

Figure 7. Efficiency of collision between a raindrop of diameter D and an aerosol particle of diameter d.

the below-cloud scavenging than to conduct experiments of in-cloud scavenging, while it is well established that the in-cloud scavenging processes are the dominant wet removal mechanisms (Flossmann et al., 1998).

The below-cloud scavenging coefficient $\lambda(d)$ has expression

$$\lambda(d) = \int_0^\infty \frac{\pi}{4} D^2 U_t(D) E_{BC}(d, D) N(D) dD \qquad (17)$$

where, $U_t(D)$ is the raindrop terminal velocity, $N(D)$ is the raindrop size distribution, and $E_{BC}(d, D)$ is the below-cloud collection efficiency. The variation of the below-cloud scavenging coefficient of particles of diameter d, given by the above equation is illustrated in Figure 8 as a function of particle size and rainfall rate, R. For a given aerosol size, we note a variation of about 2 orders of magnitude as R varies between 0.1 and 100 mm h^{-1}, which is a range that covers most of the observed precipitation rates. As discussed above, the phoretic, electric and turbulence effects are not included here. Various studies show that these effects can increase λ by one or two orders of magnitude for particles with d in the range $[0.1 - 1]$ μm (Grover et al., 1977; Tripathi and Harisson, 2002; Andronache 2004a; Chate and Devara, 2005).

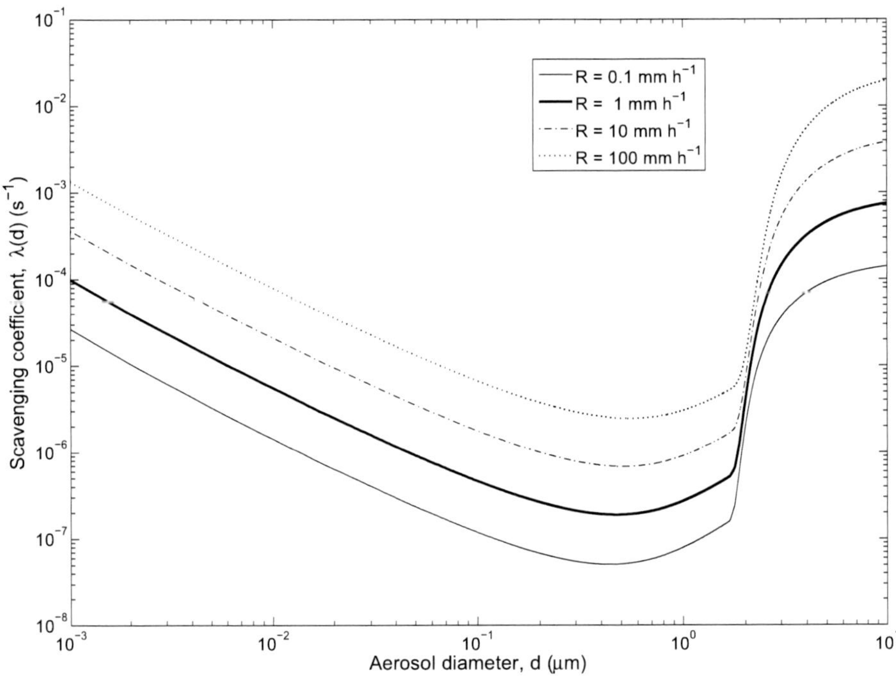

Figure 8. Scavenging coefficient λ as a function of aerosol particle d, for various rainfall intensities, R.

Larger below-cloud scavenging coefficients were determined also at the ground, by measuring the changes in aerosol size distribution during rain events (Table 3). Special attention has been given to aerosol particles with d less than about 100 nm, called ultrafine particles (UFP). Kulmala et al. (2000, 2004) showed that UFP are formed in atmosphere by nucleation of water vapor and sulfuric acid. Rain is an effective mechanism for aerosol removal from the atmosphere and it can limit the number of small aerosol particles growing to the sizes required for activation of cloud droplets. Continuous measurements over several years at a Boreal forest provided insight in the nucleation frequency, transformation and removal of UFP (Kulmala et al., 1998). One outcome of these measurements was a parameterization of the scavenging rate of UFP, λ_o from observations as described by Laakso et al. (2003):

$$log_{10}(\lambda_o/\lambda_o^{'}) = a_1 + a_2\delta^{-4} + a_3\delta^{-3} + a_4\delta^{-2} + a_5\delta^{-1} + a_6(R/R_0)^{0.5} \quad (18)$$

where: $\delta = log_{10}(d/d_0)$, $\lambda_o^{'} = 1\ s^{-1}$, $R_0 = 1$ mm h^{-1}, $d_0 = 1$ m. The coefficients $a_1, a_2, a_3, a_4, a_5, a_6$ are fitting parameters: $a_1 = 274.35758$, $a_2 = 332839.59273$, $a_3 = 226656.57259$, $a_4 = 58005.91340$, $a_5 = 6588.38582$, $a_6 = 0.244984$. It is notable that these λ_o values are generally higher than those obtained from the below-cloud scavenging model, and differences have been attributed to other factors influencing the size distribution change during rain as observed at ground (mixing between BL and cloud re-

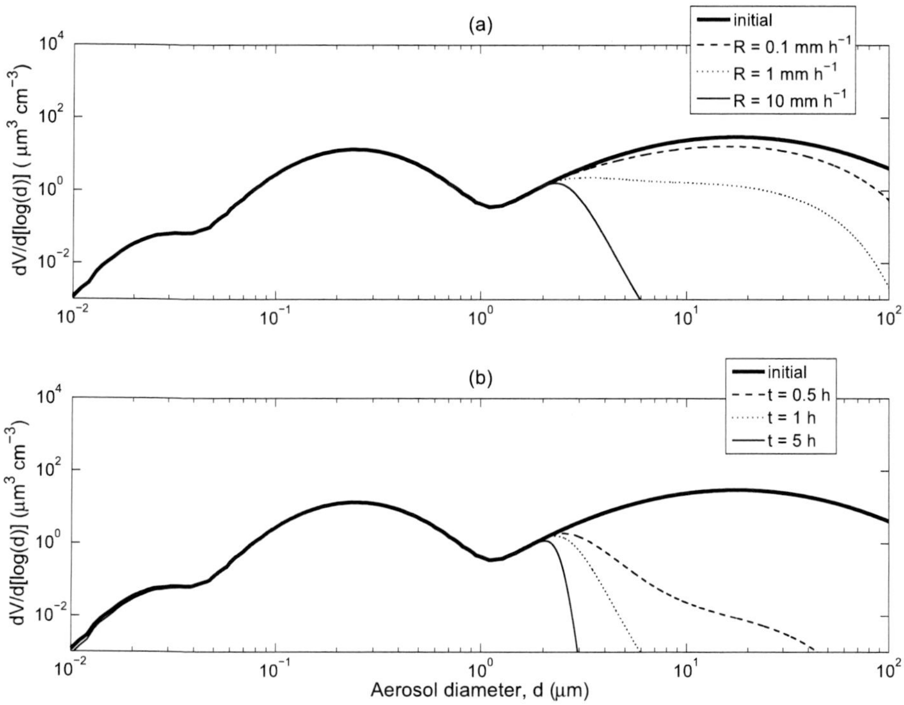

Figure 9. (a) Calculated changes in the aerosol volume distribution after one hour of rain with a rainfall rate R = 0.1 mm h^{-1}, 1 mm h^{-1} and 10 mm h^{-1}; (b) Calculated changes in the aerosol volume distribution for R = 10 mm h^{-1} after a duration of rain t = 0.5 h, 1 h and 5 h. The initial aerosol volume distribution is marked by a solid, thick line. Shown changes are due to below-cloud scavenging process only.

gion, condensation, effects of non homogeneity in horizontal direction, etc.). For details, see Andronache et al. (2006). The results by Laakso et al. (2003) appear in agreement with other reported λ_o, estimated from measurements at ground, during rain events (Table 3).

Finally, we illustrate changes in the aerosol size distribution due to the below-cloud scavenging process. We chose to show changes in the aerosol volume distribution $dV/dlog(d)$ because this is directly related to changes in aerosol mass distribution. The initial aerosol volume distribution corresponds to a remote continental type (Jaenicke, 1993). Figure 9a shows the change of aerosol volume distribution after one hour of rain with a rainfall rate R = 0.1 mm h^{-1}, 1 mm h^{-1} and 10 mm h^{-1}. The initial aerosol volume distribution is marked by a solid line and we note that after one hour, for moderate rain of intensity R = 10 mm h^{-1}, the changes are dramatic (most of the coarse particle volume is removed). Figure 9b illustrates the changes in the aerosol volume distribution for R = 10 mm h^{-1} after a duration of rain t = 0.5 h, 1 h and 5 h. We note that after 0.5 and 1 hour of rain, the coarse aerosol volume is depleted. In these plots only the below-cloud scavenging was considered, while in situations encountered during field measurements, the aerosol size distribution could change due to other interactions not included here (for a discussion see Andronache et al., 2006).

Table 3. Range of reported aerosol scavenging coefficient determined from measurements near ground during rain events.

$\lambda_o(\mathrm{s}^{-1})$	$R(\mathrm{mm\ h}^{-1})$	Location	Reference
$[0.2-1]\times10^{-4}$	1 - 3	Lexington, USA	Davenport and Peters (1978)[a]
$[0.5-4]\times10^{-4}$	0.1 - 2.6	Oxon, UK	Nicholson et al. (1991)[b]
$[0.07-4]\times10^{-4}$	0.5 - 10	Hyytiälä, Finland	Laakso et al. (2003)[c]
$[0.1-8]\times10^{-4}$	1 - 46	Pune, India	Chate and Pranesha (2004)[d]
$[0.4-1]\times10^{-4}$	1 - 20	Princeton, USA	Maria and Russell (2005)[e]
1.64×10^{-5}		Hyderabad, India	Latha et al., (2005)[f]

[a]Experiments were conducted during three rain events on September 1, October 25 and October 30, 1976. The rain events were widespread precipitation. Aerosol measurements were limited to two subranges of the size distribution: one with d in the range [7 - 22] nm, and the other with $d \geq 400$ nm.
[b]Thirty field measurements of below-cloud scavenging have been carried out in Oxon, United Kingdom, during precipitation events in 1988 and 1989. Four particle sizes (number mean diameters: 4.3 μm, 9 μm, 15 μm and 19 μm were released during precipitation. Results given above are for particles with diameter of 4.3 μm. For larger particles, the scavenging rate is increasing and authors discuss the possibility of electric charge presence, which might explain the estimated collection efficiency greater than unity.
[c]Measurements were taken during May - October months, 1996-2001. Aerosol size distribution measured for d in the range [3 - 510] nm. Rain events were characterized by predominant low intensities.
[d]Measurements were taken during a total of 17 rain events: two days in the pre-monsoon season in April 1997, and ten days post-monsoon during October and November 1997. Eight aerosol particle sizes with d in the range [13 - 750] nm were used in the experiment. Precipitations were dominated by strong convective activity and thunderstorms.
[e]Observations were conducted during 10-day period in August, 2003. Particle sizes measured ranged between 10 nm and 20 μm. Precipitation rate varied between 1 and 20 mm h^{-1}.
[f]Experiments were conducted in Hyderabad, India (17° $28'$ N and 78° $26'$ E). The estimated scavenging coefficient is representative for area average scavenging of black carbon particles in urban area.

3.2. In-cloud Scavenging

In-cloud scavenging is the main process of wet removal and is the result of nucleation scavenging of CCN, followed by rapid growth of cloud droplets and collection by falling raindrops. Contributions to in-cloud scavenging are made by collection of falling raindrops with cloud droplets (typical diameter of about 10 μm), and with interstitial aerosol (with typical d less than one micrometer). Cloud droplets collect also interstitial aerosols by coagulation. From all these processes, the collection of cloud droplets by falling raindrops is the dominant contributor to in-cloud scavenging rate, and the other processes are often neglected due to lack of information about interstitial aerosol and cloud micro-structure. The general expression for the scavenging coefficient due to in-cloud collection by falling raindrops is

$$\lambda = \int_0^\infty \frac{\pi}{4} D^2 U_t(D) E_{IC} N(D) dD \tag{19}$$

where E_{IC} is the in-cloud collection efficiency between cloud droplets (of diameter 10 μm) and raindrops. In this sense, λ is not dependent on aerosol particle diameter. E_{IC} is in the range 0.5 – 0.8 for soluble aerosol scavenged by liquid drops. For snow, the collection efficiency, $E_{IC} \sim 0.2 - 0.3$ (Scott, 1982). Scott considered a typical particle diameter inside cloud of 10 μm (after the growth of aerosol from activated aerosol particle to a cloud droplet size) and selected a value $E_{IC} = 0.65$, which corresponds to soluble aerosols such as sulfate aerosols. Following Scott (1982), d is assumed 10 μm, and for rain, λ_{IC} becomes

$$\lambda = 3.5 \times 10^{-4} R^{0.78} \tag{20}$$

where λ is in units of s^{-1} and R is in units of mm h^{-1}.

λ estimations using the model proposed by Scott are in general good agreement with values determined from field studies (Okita et al., 1996; Jylhä, 1991; Andronache 2004b) while significant variability was observed from case to case. Such variability was attributed to changes in aerosol solubility, precipitation type, vertical distribution of the precipitating system and raindrop size distribution. Nevertheless, the model proposed by Scott (1982) provides a reliable framework for calculations of λ for soluble aerosols. Table 4 shows λ values reported in literature. For air quality studies, it is useful to have a simple expression for the in-cloud scavenging, and have the rainfall intensity R prescribed from observations or forecast with a weather model. The challenge remains to validate the models against detailed surface deposition data such as those available from the National Atmospheric Deposition Program/National Trends Network (NADP/NTN) (Hicks, 2005). Additional work is needed to characterize in-cloud scavenging dependence on aerosol chemical composition and precipitation type, variation with altitude and dependence on other microphysical parameters (Svenningsson et al., 1997; Swietlicki et al., 1999; Martinsson et al., 1999).

4. Conclusion

We presented the main factors affecting the below-cloud and in-cloud scavenging coefficients of aerosols by rainfall, useful in air quality studies, and particularly in acid rain modeling. The method employed is based on explicit calculations of the efficiency of collision between a raindrop and aerosol particles. Results show that below-cloud scavenging coefficient depends mainly on the aerosol size distribution parameters and on rainfall intensity. There is evidence that:

(a) For a given aerosol size, the below-cloud scavenging coefficient varies about two orders of magnitude for a variation of rainfall rate between 0.01 and 100 mm h^{-1}. For a given rainfall rate, below-cloud scavenging coefficient varies significantly with aerosol diameter.

(b) The below-cloud scavenging is effective for very small particles (with diameters less than 0.01 μm) and for coarse particles (with diameters larger than 2 μm). For these cases,

Table 4. Reported in-cloud scavenging rates, $\lambda = a\,R^{\,b}$ (in units s^{-1}), with rainfall intensity R in mm h^{-1} units.

a	b	Reference
3.5×10^{-4}	0.78	Scott (1982)[a]
5.0×10^{-5}	0.80	Raes (1991)
8.4×10^{-5} - 3.4×10^{-4}	0.79	Maryon et al. (1992)
1.0×10^{-4}	0.67 - 0.76	Okita et al. (1996) [b]
2.3×10^{-5} - 6.6×10^{-4}	0.80	Mircea and Stefan (1998)[c]
8.0×10^{-5}	0.80	Päsler-Sauer (2003)
2.5×10^{-4} - 6.9×10^{-4}	0.68 - 0.93	Andronache (2004b)[d]
4.2×10^{-4}	0.79	ENVIRON (2005)

[a]The scavenging coefficient for rain was obtained using a Marshall and Palmer DSD, and assuming $E = 0.65$. For snow, Scott found $\lambda = 2.44 \times 10^{-4}R$ (s^{-1}), assuming $E = 0.1$.
[b]Measurements of SO_2 and SO_4^{2-} were conducted between 28 January and 3 February 1992 at Aikawa on Sado Island on the Sea of Japan coast, during the winter monsoon season. The scavenging coefficient was determined from $\lambda = \frac{S_r R}{h}$, where: S_r - scavenging ratio; R - rainfall intensity; and h - mixing layer height, under the assumption of a uniform mixed layer over the sea warm current along the Sea of Japan coast. The scavenging ratio $S_r = C_w/C_a \times 10^6$, with C_w is the chemical species concentration (mg l^{-1}) in precipitation, and C_a is the chemical species concentration (μg m^{-3}) in air.
[c]For a range of DSD and rainfall types, expressions in the form $\lambda = \alpha E R^{\,\beta}$ have been determined. For the values included in table, $E = 0.1$.
[d]Average λ values for location in the Eastern United States, were estimated based on sulfate mass balance in the atmospheric domain of interest. Data used are taken from several observational networks: (1) the Atmospheric Integrated Research Monitoring Network (AIRMoN) which is part of the National Atmospheric Deposition Program/National Trends Network (NADP/NTN); (2) the Interagency Monitoring of Protected Visibility Environments (IMPROVE); and (3) the National Climatic Data Center (NCDC).

the below-cloud scavenging coefficients have values comparable with in-cloud scavenging coefficients. For rainfall rate $R \approx 1$ mm h^{-1}, the 0.5-folding time of these particles due to below-cloud scavenging is of the order of one hour. This suggests that for a model description of detailed rapid changes in aerosol size distribution in the BL, it is necessary to use bin defined below-cloud scavenging coefficients, $\lambda(d)$, and high frequency values of rainfall rate. The presented results show that the BL aerosol size distribution with coarse mode is affected by below-cloud scavenging very shortly after rain starts (in a fraction of one hour). After the removal of very small and coarse particles, the below-cloud aerosol size distribution is dominated by particles in the accumulation mode. Literature review suggests that scavenging of particles in the accumulation mode is likely enhanced by phoretic, electric and turbulence effects which are poorly represented in current models.

(c) For low and moderate rainfall rates ($R \approx [0.1 - 10]$ mm h^{-1}), the below-cloud scavenging is negligible for particles in the range $[0.1 - 1]\mu m$ in comparison with in-cloud scavenging . For very intense precipitation (R = 100 mm h^{-1}), below-cloud scavenging of these particles becomes significant. These results have practical significance for modeling rapid changes of aerosol size distribution in the BL as in the cases of field experiments with continuous high frequency measurements of aerosol size spectrum. Thus, in case of precipitation in the measurement domain, the below-cloud scavenging can contribute substantially to the removal of fine and coarse particles. Contribution to the acid rain at ground from such particles can be due to acidic ultrafine particles as well as from sulphate deposited on coarse particles (such as mineral and combustion aerosol).

(d) In-cloud aerosol scavenging is the dominant wet removal process and for practical purposes it can be approximated by an expression in the form $\lambda = a\ R^{\,b}$ (in units s^{-1}), with rainfall intensity R in mm h^{-1} units. Coefficients a and b show variations in the published literature and they reflect the variability in the aerosol and cloud microphysics, including the collection efficiency E between raindrops and cloud droplets and aerosols in-cloud. This simple expression for λ shows also the importance of correct assessment of the rainfall intensity R, which remains a challenging task for weather modeling and observation systems. Recent developments in high resolution weather modeling, such as using the Weather Research and Forecasting (WRF) Model (Skamarock et al., 2005), and specialized air quality models, as well as in the radar technology used in precipitation measurements provide new venues of progress in the area of aerosol wet removal and acid rain simulations.

Work is needed to improve the accuracy of scavenging coefficients used in air quality and acid rain problems. Some of the outstanding issues which require further evaluation are: 1) determine the role of phoretic, electric forces and turbulence on the collection efficiency between raindrop and aerosols; 2) extend aerosol scavenging studies to snow and ice scavenging and provide robust parameterizations of these processes; 3) evaluate scavenging schemes in comprehensive laboratory studies, covering a wide range of conditions as encountered in real atmosphere; 4) evaluate the scavenging schemes in well designed field experiments; 5) evaluate wet scavenging schemes in air quality models using surface measurements related to acid rain as those available form acid deposition programs.

Acknowledgments

The author is grateful to Charles Seman, Stephen Jascourt, and Vaughan Phillips for valuable discussions and information on cloud dynamics, rainfall measurements and aerosol cloud-interactions. Many thanks to Lauri Laakso and Tiia Grönholm for the opportunity to work with aerosol measurements data from the University of Helsinki. The author also thanks Jan Peter Muller, Robin Hogan, and Marion Mittermaier for kindly providing information on cloud vertical structure observations. The author acknowledges the use of data from the National Climatic Data Center (NCDC), from the Interagency Monitoring of Protected Visibility Environments (IMPROVE) and from the Atmospheric Integrated Research Monitoring Network (AIRMoN). The AIRMoN program is part of the NADP/NTN.

5. Notations

d	aerosol particle diameter
D	raindrop diameter
E	collection efficiency
D_a	aerosol diffusivity in air
k_B	Boltzmann constant
$n(d)$	aerosol size distribution
$N(D)$	raindrop size distribution (DSD)
N_0	Marshall and Palmer DSD parameter
N_{0g}	Gamma DSD parameter
P	atmospheric pressure
R	rainfall rate
R_e	raindrop Reynolds number
S_c	aerosol Schmidt number
W_e	raindrop Weber number
U_t	raindrop terminal velocity
t_c	air temperature in Celsius degrees
T_a	absolute temperature of air
λ	scavenging coefficient
μ_a	air viscosity
μ_g	Gamma DSD parameter
ρ_a	density of air
ρ_p	aerosol particle density
ρ_w	raindrop water density
σ	raindrop surface tension
Λ	Marshall and Palmer DSD parameter
Λ_g	Gamma DSD parameter

References

[1] Adams, P. J., and Seinfeld, J. H., Predicting global aerosol size distributions in general circulation models, *J. Geophys. Res.,* **107**, 4370, doi : $10.1029/2001JD001010$, 2002.

[2] Andronache, C., Estimated variability of below-cloud aerosol removal by rainfall for observed aerosol size distributions, *Atmos. Chem. Phys.*, **3**, 131–143, 2003.

[3] Andronache, C.: Diffusion and electric charge contributions to below-cloud wet removal of atmospheric ultra-fine aerosol particles, *J. Aerosol Sci.,* **35**, 1467–1482, 2004a.

[4] Andronache, C., Estimates of sulfate aerosol wet scavenging coefficient for locations in the Eastern United States, *Atmos. Environ.* **38**, 795–804, 2004b.

[5] Andronache, C., T. Grönholm, L. Laakso, V. Phillips, and A. Venäläinen, Scavenging of ultrafine particles by rainfall at a boreal site: observations and model estimations, *Atmos. Chem. Phys.,* **6**, 4739-4754, 2006.

[6] Atlas, D. and Ulbrich, C. W., Path and area integrated rainfall measurement by microwave attenuation in the 1–3 cm band, *J. Appl. Meteorol.*, **16**, 1322–1331, 1977.

[7] Bae, Soo Ya, Chang Hoon Jung, and Yong Pyo Kim, Development and evaluation of an expression for polydisperse particle scavenging coefficient for the below-cloud scavenging as a function of rain intensity using the moment method, *J. Aerosol Sci.*, **37**, 1507–1519, 2006.

[8] Beard, K. V., Simple altitude adjustments to raindrop velocities for Doppler radar analysis, *J. Atmos. Sci.*, Vol. 2, 468 – 471, 1985.

[9] Beard, K. V., R. J. Kubesh, and H. T. Ochs, Laboratory measurements of small raindrop distorsion, Part I: Axis ratios and fall behaviour, *J. Atmos. Sci.*, **48**, 698 – 770, 1991.

[10] Bennett, J. A., D. J. Fang, and R. C. Boston, The relationship between N_0 and Λ for Marshall - Palmer type raindrop-size distributions, *J. Clim. and Appl. Meteo,* Vol. 23, 768–771, 1984.

[11] Best, A. C., Empirical formulae for the terminal velocity of water drops falling through the atmosphere, *Quart. J. Royal. Meteor. Soc.*, **76**, 302–311, 1950.

[12] Byrne, M. A. and Jennings, S. G., Scavenging of sub-micrometer aerosol particles by water drops, *Atmos. Environ.,* **27A**, 14, 2099–2105, 1993.

[13] Carslaw, K. S., Harrison, R. G., and Kirkby, J., Cosmic rays, clouds and climate, *Science*, **298**, 1732-1737, 2002.

[14] Chate, D. M. and Pranesha, T. S., Field studies of scavenging of aerosols by rain events, *J. Aerosol Sci.*, **35**, 695–706, 2004.

[15] Chate, D. M, and P. C. S. Devara, Parametric study of scavenging of atmospheric aerosols of various chemical species during thunderstorm and non thunderstorm rain Events, *J. Gephys. Res.*, Vol. 110, D23208, doi : $10.1029/2005JD006406$, 2005.

[16] Dana, M. T. and Hales, J. M., Statistical aspects of the washout of polydisperse aerosols, *Atmos. Environ.*, **10**, 45–50, 1976.

[17] Davenport, H. M. and Peters, L. K., Field studies of atmospheric particulate concentration changes during precipitation, *Atmos. Environ.*, **12**, 997–1008, 1978.

[18] Debry, E., K. Fahey, K. Sartelet, B. Sportisse, and M. Tombette, Technical Note: A new SIze REsolved Aerosol Model (SIREAM), *Atmos. Chem. Phys.,* **7**, 1537-1547, 2007.

[19] Dingle, A. N., and Y. Lee, Terminal fallspeeds of raindrops, *J. Applied. Meteor.*, Vol. 11, 877–879, 1972.

[20] ENVIRON, Users Guide CAMxVersion 4.20. Technical report, ENVIRON International Corporation, 2005.

[21] Feng, J., A 3-mode parameterization of below-cloud scavenging of aerosols for use in atmospheric dispersion models, *Atmos. Environ.*, **41**, 6808-6822, 2007.

[22] Flossmann, A. I., W. D. Hall, and H. R. Pruppacher, A theoretical study of the wet removal of atmospheric pollutants. Part I: The redistribution of aerosol particles captured through nucleation and impaction scavenging by growing cloud drops, *J. Atmos. Sci.*, Vol. 42, No. 4, 583–606, 1985.

[23] Flossmann, A. I., H. R. Pruppacher, and J. H. Topalian, A theoretical study of the wet removal of atmospheric pollutants. Part II: The uptake and redistribution of (NH4)2SO4 particles and SO2 gas simultaneously scavenged by growing cloud drops, *J. Atmos. Sci.*, Vol. 44, No. 20, 2912–2923, 1987.

[24] Flossmann, A., Interaction of aerosol particles and clouds, *J. Atmos. Sci.*, **65**, 879–887, 1998.

[25] Foote, G. B., and P. S. du Toit, Terminal velocity of raindrops aloft, *J. Appl. Meteor.*, **8**, 249 – 253, 1969.

[26] Garcia Nieto, P. J., B. Arganza Garcia, J. M. Fernandez Diaz and M. A. Rodriguez Brana, Parametric study of selective removal of atmospheric aerosol by below-cloud scavenging, *Atmos. Environ.*, Vol. 28, No. 14, 2335–2342, 1994.

[27] Ghan, S. J., and Schwartz, S. E., Aerosol properties and processes: A path from field and laboratory measurements to global climate models, *Bull. Amer. Meteor. Soc.*, Vol. 88, No. 7, pp. 1059–1084, 2007.

[28] Gong, S. L., Barrie, L. A., Blanchet, J.-P., v. Salzen, K., Lohmann, U., Lesins, G., Spacek, L., Zhang, L. M., Girard, E., Lin, H., Leaitch, R., Leighton, H., Chylek, P., and Huang, P., Canadian Aerosol Module: A size-segregated simulation of atmospheric aerosol processes for climate and air quality models, 1, Module development, *J. Geophys. Res.*, **108**(D1), 4007, doi : $10.1029/2001JD002002$, 2003.

[29] Gossard, E. E., R. G. Strauch, D. C. Welsh, and S. Y. Matrasov, Cloud layer, particle identification and rain rate profiles from ZRV_f measurements by clean-air Doppler radars, *J. Atmos. Oceanic Technol.*, **9**, 108–119, 1992.

[30] Greenfield, S., Rain scavenging of radioactive particulate matter from the atmosphere, *J. Meteorol.*, **14**, 115–125, 1957.

[31] Grell, G. A., S. E. Peckham, R. Schmitz, S. A. McKeen, G. Frost, W. C. Skamarock, and B. Edere, Fully coupled online chemistry within the WRF model, *Atmos. Environ.*, **39**, 6957-6975, 2005.

[32] Grover, S. N., Pruppacher, H. R., and Hamielec, A. E., A numerical determination of the efficiency with which spherical aerosol particles collide with spherical water

drops due to inertial impaction and phoretic and electric forces, *J. Atmos. Sci.,* **34**, 1655–1663, 1977.

[33] Gunn, R., and G. D. Kinzer, The terminal velocity of fall for water droplets in stagnant air, *J. Meteor.*, **8**, 249 – 253, 1949.

[34] Harrison, R. G. and Carslaw, K. S., Ion-aerosol-cloud processes in the lower atmosphere, *Rev. Geophys.*, **41**(3), 1012, *doi* : 10.1029/2002*RG*000114, 2003.

[35] Henzing, J. S., D. J. L. Olivie, and P. F. J. van Velthoven, A parameterization of size resolved below cloud scavenging of aerosols by rain, *Atmos. Chem. Phys.,* **6**, 3363–3375, 2006.

[36] Hicks, B., A climatology of wet deposition scavenging ratios for the United States *Atmos. Environ.*, Vol. 39, 9, 1585–1596, 2005.

[37] Jacobson, M. Z., Analysis of aerosol interactions with numerical techniques for solving coagulation, nucleation, condensation, dissolution, and reversible chemistry among multiple size distributions, *J. Geophys. Res.,* **107**(D19), 4366, *doi* : 10.1029/2001*JD*002044, 2002.

[38] Jaenicke, R., Tropospheric aerosols, In *Aerosol-cloud-climate interactions*, Peter Hobbs (Ed), Academic Press, San Diego, CA, pp. 1–31, 1993.

[39] Jennings, S. G., Wet processes affecting atmospheric aerosols, in *Atmospheric Particles,* Edited by R. M. Harrison and R. Van Grieken, John Wiley and Sons Ltd., pp. 601, 1998.

[40] Joss, J., J. C. Thams and A. Waldvogel, The variation of raindrop-size distributions at Lacarno. Proc. Int. Conf. Cloud Physics, Toronto, *Amer. Meteor. Soc.*, 369 – 373, 1968.

[41] Jung C.H., Kim Y.P., and Lee K.W., A moment model for simulating raindrop scavenging of aerosols, *J. Aerosol Sci.*, Vol. 34, 9, pp. 1217–1233, 2003.

[42] Jylhä, K., Empirical scavenging coefficients of radioactive substances released from Chernobyl, *Atmos. Environ.*, **25A**, 263–270, 1991.

[43] Kessler, E., On the distribution and continuity of water substance in atmospheric circulations, *Meteorol. Monographs* No. **32**, 84 pp., *Am. Meteorol. Soc.,* Boston, Mass., 1969.

[44] Kulmala, M., Toivonen, A., Mäkelä, J. M., and Laaksonen, A., Analysis of the growth of nucleation mode particles observed in Boreal forest, *Tellus*, **50B**, 449–462, 1998.

[45] Kulmala, M., Pirjola, L., and Mäkelä, M., Stable sulfate clusters as sources of new atmospheric particles, *Nature*, **404**, 66–69, 2000.

[46] Kulmala, M., Vehkamäki, H., Petäjä, T., Dal Maso, M., Lauri, A. Kerminen, V.-M., Birmili, W., and McMurry, P., Formation and growth rates of ultrafine atmospheric particles: a review of observations, *J. Aerosol Sci.,* **35**, 143–176, 2004.

[47] Laakso, L., Grönholm, T., Rannik, Ü., Kosmale, M., Fiedler, V., Vehkamäki, H., and Kulmala, M., Ultrafine particle scavenging coefficients calculated from 6 years field measurements, Atmos. Environ., **37**, 3605–3613, 2003.

[48] Latha, K. M., K. V. S. Badarinath, and P. M. Reddy, Scavenging efficiency of a rainfall on black carbon aerosols over an urban environment, *Atmos. Sci. Let.*, **6**: 148–151, 2005.

[49] Laws, J,. O, and D. A. Parsons, The relation of raindrop-size to intensity, *Trans. Amer. Geophys. Union,* **24**, Part II, 452–460, 1943.

[50] Lohmann U., and J. Feichter, Global indirect aerosol effects: a review, *Atmos. Chem. Phys.*, 5, 715–737, 2005.

[51] Loosmore, G.A., and Cederwall, R.T., Precipitation scavenging of atmospheric aerosols for emergency response applications: testing an updated model with new real-time data, *Atmos. Environ.* **38**, 993-1003, 2004.

[52] Maria, S. S., and Russell, L. M., Organic and Inorganic Aerosol Below-Cloud Scavenging by Suburban New Jersey Precipitation, *Environ. Sci. Technol.,* **39**(13) 4793–4800, 2005.

[53] Marshall, J. S., and Palmer, W. M., The distribution of raindrop with size, *J. Meteorol.*, **5**, 165–166, 1948.

[54] Martinsson, B. G., Frank, G., Cederfelt, S. I., et al., Droplet nucleation and growth in orographic clouds in relation to aerosol population, *Atmos. Res.,* **50**, 289–315, 1999.

[55] Maryon, R.H., Smith, F.B., Conwy, B.J., Goddard, D.M., The UK nuclear accident response model (NAME), *Progress in Nuclear Energy* **26** (2), 85-104, 1992.

[56] McDonald, J. E., The shape and aerodynamics of large droplets, *J. Meteor. Soc.,* **11**, 478 – 494, 1954.

[57] McGann, B. T., and Jennings, S. G., The efficiency with which drizzle and precipitation sized drops collide with aerosol particles, *Atmos. Environ.,* **25A**, 3/4, 791–799, 1991.

[58] Mircea, M., and Stefan, S., A theoretical study of the microphysical parameterization of the scavenging coefficients as function of precipitation type and rate, *Atmos. Environ.* **32** (17), 2931-2938, 1998.

[59] Mircea, M., S. Stefan, and S. Fuzzi, Precipitation scavenging coefficient: influence of measured aerosol and raindrop size distributions, *Atmos. Environ.,* **34**, 5169–5174, 2000.

[60] Mueller, S. F., E. M. Bailey, T. M. Cook, and Q. Mao, Treatment of clouds and the associated response of atmospheric sulfur in the Community Multiscale Air Quality (CMAQ) modeling system, *Atmos. Environ.*, **40**, 6804-6820, 2006.

[61] Nicholson, K. W., J. R. Branson, and P. Giess, Field measurements of the below-cloud scavenging of particulate material, *Atmos. Environ.,* **25A**, 3/4, 771–777, 1991.

[62] Okita, T., H. Hara, and N. Fukuzaki, Measurements of atmospheric SO2 and SO4, and determination of the wet scavenging of sulfate aerosol for the winter monsoon season over the sea of Japan, *Atmos. Environ.*, Vol. 30, No. 22, pp. 3733–3739, 1996.

[63] Päsler-Sauer, J., Description of the atmospheric dispersion model ATSTEP. Version RODOS PV 5.0. Technical Report RODOS(RA2)-TN(03)-01, Forschungszentrum Karlsruhe GMBH, 2003.

[64] Pruppacher, H. R. and Klett, J. D., *Microphysics of clouds and precipitation,* Kluwer Academic Publishers, Dordrecht, Boston, London, pp. 954, 1998.

[65] Radke, L. F., P. V. Hobbs, and M. W. Eltgroth, Scavenging of aerosol particles by precipitation, *J. Atmos. Sci.*, Vol. 19, 715–722, 1980.

[66] Raes, F., Tassone, C., Grippa, G., Zarimpas, N., Graziani, G., Updating long-range transport model predictions using real-time monitoring data in case of nuclear accidents with release to the atmosphere, *Atmos. Environ.* **25A** (12), 28072814, 1991.

[67] Ramanathan, V., P. J. Crutzen, J. T. Kiehl, and D. Rosenfeld, Aerosols, climate, and the hydrological cycle, *Science*, **7**, Vol. 294, No. 5549, pp. 2119–2124, 2001.

[68] Rasch, P. J., Feichter, J., Law, K., et al., A comparison of scavenging and deposition processes in global models: results from the WCRP Cambridge Workshop of 1995, *Tellus*, **52B**, 1025–1056, 2000.

[69] Schumann, T., Aerosol and hydrometeor concentrations and their chemical composition during winter precipitation along mountain slope – III. Size-differentiated in-cloud scavenging efficiencies, *Atmos. Environ.*, **25A**, 3/4, 809–824, 1991.

[70] Scott, B. C., Theoretical estimates of the scavenging coefficient for soluble aerosol particles as a function of precipitation type, rate, and altitude, *Atmos. Environ.*, **16**, 7, 1753–1762, 1982.

[71] Sehmel, G.A., Particle and gas deposition: a review, *Atmos. Environ.*, **14**, 983-1011, 1980.

[72] Seinfeld, J. H. and Pandis, S. N., *Atmospheric Chemistry and Physics*, Wiley, New York, 1326 pp., 1998.

[73] Sekhon, R. S., and R. C. Srivastva, Doppler radar observations of drop-size distributions in a thunderstorm, *J. Atmos. Sci.*, Vol. 28, 983–994, 1971.

[74] Skamarock, W. C., J. B. Klemp, J. Dudhia, D. O. Gill, D. M. Barker, W. Wang, J. G. Powers, A description of the Advanced Research WRF Version 2. *NCAR Tech. Note,* ncar/TN-468+STR, 88 pp., 2005

[75] Slinn, W. G. N., and Hales, J. M., A reevaluation of the role of thermophoresis as a mechanism of in- and below-cloud scavenging, *J. Atmos. Sci.*, **28**, 1465–1471, 1971.

[76] Slinn, W. G. N., Some approximations for the wet and dry removal of particles and gases from the atmosphere, *Water, Air, and Soil Pollut.*, **7**, 513–543, 1977.

[77] Slinn, W. G. N., Precipitation scavenging, in: *Atmospheric Sciences and Power Production* – 1979, chap. 11, Division of Biomedical Environmental Research, U.S. Department of Energy, Washington, D.C., 1983.

[78] Sparmacher, H., Fulber, K., and Bonka, H., Below-cloud scavenging of aerosol particles: Particle-bound radionuclide – *Experimental, Atmos. Environ.*, **27A**, 4, 605–618, 1993.

[79] Sportisse, B., A review of parameterizations for modelling dry deposition and scavenging of radionuclides, *Atmos. Environ.,* **41**, 2683-2698, 2007.

[80] Svenningsson, B., Hansson, H.-C., Martinsson, B., et al., Cloud droplet nucleation scavenging in relation to the size and hygroscopic behavior of aerosol particles, *Atmos. Environ.,* **31**, 16, 2463–2475, 1997.

[81] Swietlicki, E., Zhou, J., Berg, O. H., et al., A closure study of sub-micrometer aerosol particle hygroscopic behavior, *Atmos. Res.,* **50**, 205–240, 1999.

[82] Ten Brink, H. M., Schwartz, S. E., and Daum, P. H., Efficient scavenging of aerosol sulfate by liquid-water clouds, *Atmos. Environ.*, **21**, 9, 2035–2052, 1987.

[83] Tinsley, B. A., Rohrbaugh, R. P., Hei, M., and Beard, K. V., Effects of image charges on scavenging of aerosol particles by cloud droplets and on droplet charging and possible ice nucleation processes, *J. Atmos. Sci.,* **57**, 2118–2134, 2000.

[84] Tombette, M., and B. Sportisse, Aerosol modeling at a regional scale: Model-to-data comparison and sensitivity analysis over Greater Paris, *Atmos. Environ.* **41**, 6941-6950, 2007.

[85] Tost, H., Jöckel, P., Kerkweg, A., Sander, R., and Lelieveld, J., Technical note: A new comprehensive SCAVenging submodel for global atmospheric chemistry modeling, *Atmos. Chem. Phys.*, **6**, 565–574, 2006.

[86] Tripathi S.N., and Harrison R.G., Enhancement of contact nucleation by scavenging of charged aerosol, *Atmos. Res.,* **62**, 57–70, 2002.

[87] Ulbrich, C., Natural variations in the analytical form of the raindrop size distribution, *J. Clim. Appl. Meteorol.* **22**, 1764-1775, 1983.

[88] Ulbrich C. W., and D. Atlas, Rainfall microphysics and radar properties: analysis method for drop size spectra, *J. Appl. Meteo.,* Vol. 37, 912 – 923, 1998.

[89] Uplinger, W. G., *A new formula for raindrop terminal velocity,* Preprints 20*th* Radar meteorology Conference, Am. Meteor. Soc., Boston, pp. 389 – 391, 1977.

[90] Van Boxel, J. H., *Numerical model for the fall speed of raindrops in a rainfall simulator,* Workshop on wind and erosion, Ghent, Belgium, November 17 – 18, 1997.

[91] Waldvogel, A., The N_0 jump of raindrop spectra, *J. Atmos. Sci.* **31**, 1067–1078, 1974.

[92] Wang, P. K. and Pruppacher, H. R., An experimental determination of the efficiency with which aerosol particles are collected by water drops in subsaturated air, *J. Atmos. Sci.*, **34**, 1664–1669, 1977.

[93] Wang, P. K., Grover, S. N., and Pruppacher, H. R., On the effect of electric charges on the scavenging of aerosol particles by clouds and small raindrops, *J. Atmos. Sci.*, **35**, 1735–1743, 1978.

[94] Zhang, G., J. Vivekanandan, and E. Brandes, A method for estimating rain rate and drop size distribution from polarimetric radar measurements, *IEEE Transactions in Geosciences and Remote Sensing,* **39**, 830–841, 2001.

[95] Zhang, L., Michelangeli, D. V., and Taylor, P. A., Numerical studies of aerosol scavenging by low-level, warm stratiform clouds and precipitation, *Atmos. Environ.*, **38**, 4653–4665, 2004.

[96] Zhao, H., and Zheng, C., Monte Carlo solution of wet removal of aerosols by precipitation. *Atmos. Environ.* **40**, 1510-1525, 2006.

In: Acid Rain Research Focus
Editor: Lyman G. Roglesfield, pp. 75-103
ISBN: 978-1-60456-373-3

Chapter 3

LONG - TERM MONITORING OF WET AND DRY DEPOSITIONS IN A PO VALLEY CITY (BOLOGNA, ITALY)

Luciano Morselli[1], Elena Bernardi[1], Ivano Vassura[1], Fabrizio Passarini[1] and Enrico Tesini[2]
[1] Dept. of Industrial Chemistry and Materials, University of Bologna, Bologna, Italy
[2] ENEA PROT-IDR, Bologna, Italy

Abstract

Urban areas generally undergo high pollution loads, specially due to traffic and industrial plants, and particular microclimate conditions, depending on wind pattern and topographic structure. This results in higher contamination than in natural sites, with harmful and undesired effects on human health, ecosystem and materials, in particular those constituting cultural heritage. Sampling and analysis of wet and dry atmospheric deposition can provide a clear picture of the environmental quality of an ecosystem and can lead to actions aimed at reducing pollutant emissions in the atmosphere. This chapter describes the results of a nine years monitoring (1995-2003) of wet and dry depositions collected in the urban site of Bologna, with a DDAS (Dry Deposition on Aquatic Surface) sampler. Bologna is located in the Po Valley (Northern Italy), a heavily industrialized and densely populated region, characterized by a typical atmospheric circulation frequently associated with the air mass stagnation conditions.

The purposes of this study were to investigate the chemical composition of both wet and dry depositions, with regard to the main cations (H^+, Ca^{2+}, Mg^{2+}, Na^+, K^+, NH_4^+), anions (SO_4^{2-}, NO_3^-, Cl^-) and heavy metals (Cd, Cr, Cu, Ni, Pb, V, Zn), to examine the temporal trends of the main ions, to identify the contribution of different sources to the chemical composition, in particular distinguishing between natural and anthropogenic ones, and to point out local, regional and long-range influences. The study of correlation between ion pairs and data factor analysis represented a useful tool to this aim.

Three main factors were identified which affected chemical composition in the 9 years survey: anthropogenic, marine and terrigenous contributions, acting respectively on a local, regional and both local and long-range scale. The separate study of wet and dry deposition revealed a predominance of anthropogenic species fallout in the former mechanism, due to their presence in the finest particles, compared to those from natural sources. Finally, a clear decrease of free acidity in atmospheric deposition was also observed, associated to the lowering of anthropogenic species, as nitrate and sulphate ions; this suggests an improvement of air quality in Bologna urban area, due to the recent advances in pollutant control.

Introduction

Wet and dry depositions are the most important processes governing the transfer of the airborne chemicals from the atmosphere on to surfaces [1, 2]. Depending on the nature and the amount of the removed species, atmospheric depositions can be responsible for detrimental (i.e. acidification, eutrophication, accumulation of heavy metals, corrosion, monument soiling…) or positive (i.e. acidity neutralization, input of nutrient elements…) effects on ecosystems and materials, in particular those constituting cultural heritage [3, 4, 5, 6, 7, 8].

A quantitative estimation of the different vulnerability of a certain ecosystem or material can be performed by the calculation of the Critical Load or Acceptable Levels, respectively [9]. Considering the ecosystems, the Critical Load is defined as a quantitative estimate of an exposure to one or more pollutants below which significant harmful effects on specified elements of the environment do not occur, according to present knowledge [10]. The difference between actual deposition and the critical load, named exceedance, can provide a good estimation of the present risk of an area [11, 12, 13, 14]. In the last decades, the deposition fluxes of different pollutants have been widely investigated in Europe, in the framework of UN-ECE Convention on Long Range Transboundary Air Pollution, to know the extent of the problem on continental and national scales and to take the appropriate measures [15].

In fact, sampling and analysis of wet and dry atmospheric deposition are fundamental not only for determining when critical loads are exceeded, but also for identifying the main sources of the different compounds, understanding local, regional and long-range influences, validate diffusional and depositional models and for demonstrating, by the analysis of temporal trends, the efficiency of emissions reduction policies on a regional scale.

Urban areas generally undergo high pollution loads, specially due to traffic and industrial plants, and particular microclimate conditions, depending on wind pattern and topographic structure, which result in higher contamination than in natural sites [16, 17].

In particular, Bologna is located in the Po Valley (Central-Northern Italy), a densely populated region, characterized by intensive industrial, trading and agricultural activities. In this area, the atmospheric circulation is often associated with stagnant air mass conditions, in particular during wintertime, when frequent thermal inversions at low altitude, low mixing layer heights and prolonged foggy or hazy periods occur. [18, 19]

In order to investigate the chemical composition, as regards the main ions and heavy metals, to examine fluxes and temporal trends, to identify the contribution of natural or anthropogenic sources and to point out local, regional and long-range influences on atmospheric depositions in Bologna, a long term monitoring activity started in April 1995, with the analysis of the main ions. In 1998, the analysis of heavy metals integrated the monitoring activity. In this chapter, the most significant results obtained during the period 1995-2003 are discussed.

Even if in the scientific literature the study of atmospheric deposition pointed a larger attention on the chemistry of precipitation, the importance of the dry contribution has been ascertained, after an early underestimation [20]. Therefore, this monitoring was performed collecting separately wet and dry depositions [21].

The main statistical tools, used to interpret the chemical information obtained from the analyses, were the calculation of correlation between ion pairs and factor analysis.

Material and Methods

The sampling site is located in the North-East outskirts of Bologna, at about 4.5 Km from the city centre. The site, looking on to a public park, is about 1 Km far from a motorway and a bypass and about 10 Km far from both an incineration plant and Bologna airport. Different industrial and handicraft activities are in the vicinity of the site, as well as agricultural areas. Moreover, Adriatic Sea is about 80 Km east far from Bologna, and Tyrrhenian Sea about 95 Km southwest.

The deposition sampler was placed on the roof of three-storey Research Center building of ENEA (Italian National Agency for New Technologies, Energy and the Environment), following the RIDEP (Italian Network for the Study of Atmospheric Deposition) guideline for the sampler setting [21, 22].

RIDEP protocols [23, 24] and specific procedures, developed in particular for dry deposition and heavy metals [25, 26, 27], were followed for sampling, analysis, quality control and data elaboration.

In order to sample wet and dry deposition separately, a ''DDAS'' (dry deposition on aquatic surface) sampler-MTX Italia was used. The collector consists in two high-density polyethylene cylindrical vessels, equipped with an electrical rain sensor capable of triggering the cover so as to protect the dry sample, in case of precipitation, and to collect the wet deposition in the other container. The wet container has a surface of 0.0661 m^2, while the dry vessel exposes to the atmosphere a constant layer of ultrapure water (R>18 ΩM—Milli-Q system) 0.0531 m^2 wide and 1.5 cm thick. The constancy of the level is assured through a level sensor connected to an automatic pump drawing in an ultrapure water tank. The water surface is often used to sample dry deposition, due to its advantage with respect to solid surfaces: better efficiency, minor resuspension, better representativeness of natural surfaces [2, 16, 25, 28].

Sampling was carried out every week for wet deposition and every two weeks for dry one.

For each sample, volume, pH, alkalinity and conductivity were measured immediately after the sampling. Then, samples were filtered, in order to separate soluble and insoluble fractions. On the first fractions, main cations (Ca^{2+}, Mg^{2+}, Na^+, K^+, NH_4^+), anions (SO_4^{2-}, NO_3^-, Cl^-) and metals (Cd, Cr, Cu, Ni, Pb, V, Zn) were analysed. Considering the insoluble fraction, filters were mineralised and the same metals as in the soluble portion analysed. The experimental procedure (sampling, treatment, storage, analysis) is summarized in Figure 1.

For cation and anion analysis, an internal quality control was executed, following the RIDEP procedures: on the basis of ionic and conductivity balances, results were classified in 3 classes and accepted if less than 50% fell in the third class.

Whilst, from 1995 to 1999, results in class 3 were on average 19% for dry samples and 13% for wet ones [29], from 2000 to 2003 the average percentage resulted respectively 5% and 4%.

For metal analysis, the digestion step and the whole procedure were tested using a certified reference material (CRM; NIST 1648—urban particulate matter) [26, 27]. Moreover, the station laboratory participated to periodical comparison exercises among network laboratories.

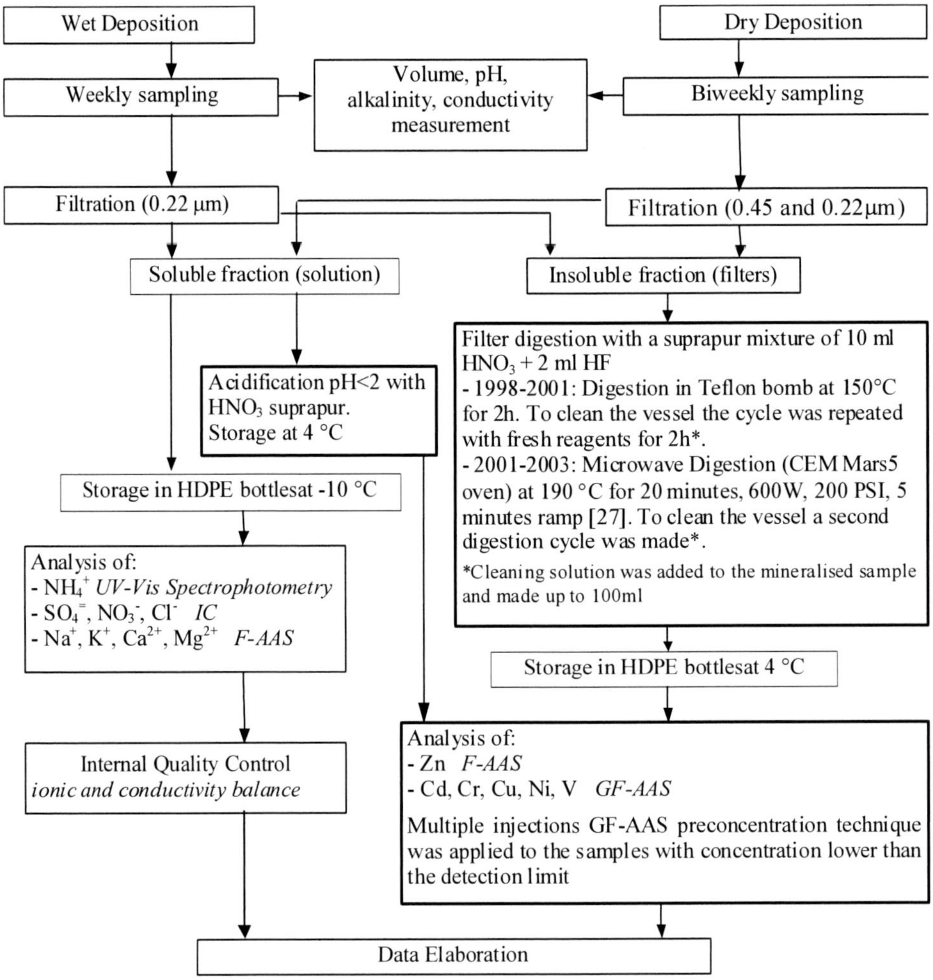

Figure 1. Scheme of the experimental procedure.

Besides the calculation of deposition fluxes and trends, analytical data were statistically elaborated, in particular through correlation between ion pairs [26, 30] and data factor analysis. The calculation of Principal Components and the following Factor Analysis (by Varimax rotation) was performed using Q-Parvus 3.0 software package [31].

Results and Discussion

Deposition Fluxes and Temporal Trends

Considering, at first, the annual deposition fluxes for the main ions (Figure 2) and metals (Figure 3), it has to be noticed that the main contribution to the total fallout of most species is provided

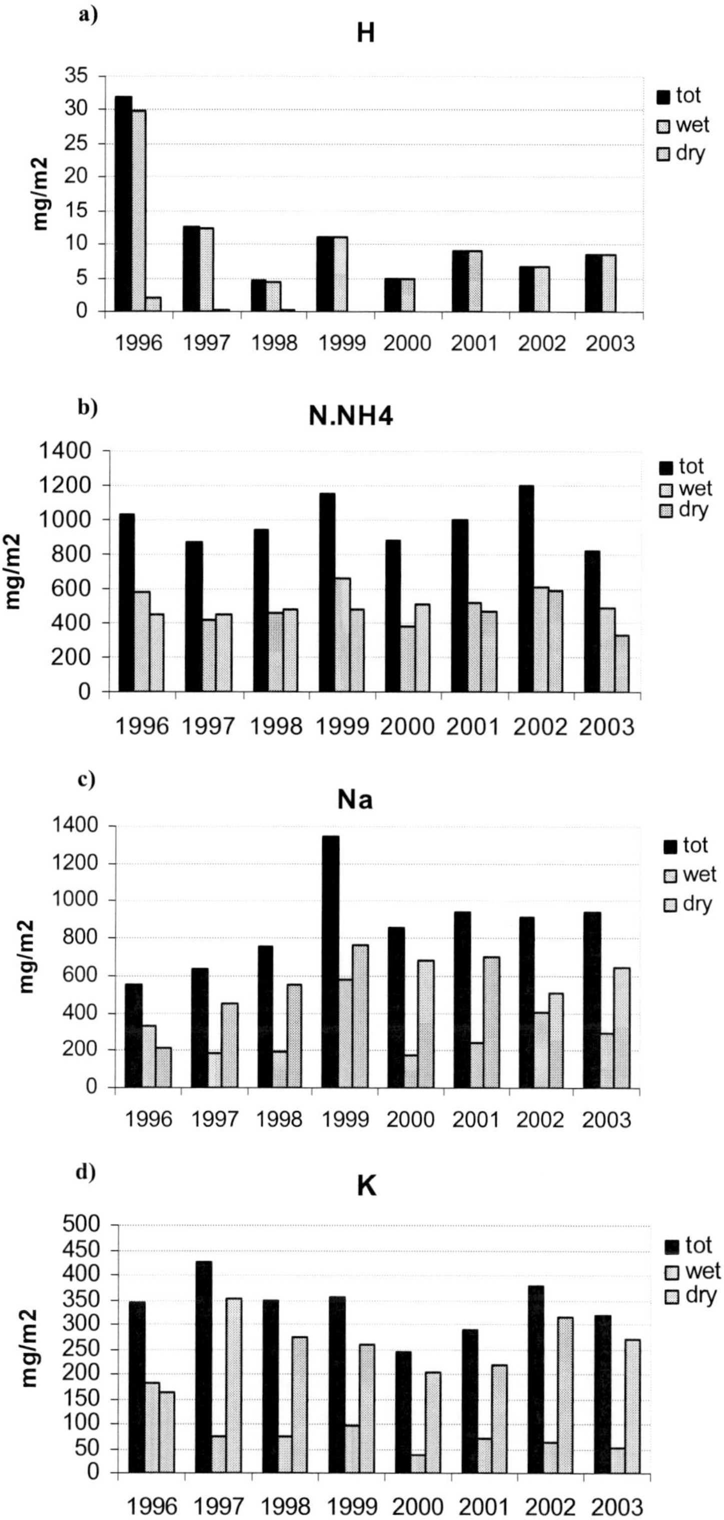

Figure 2. Continued on next page.

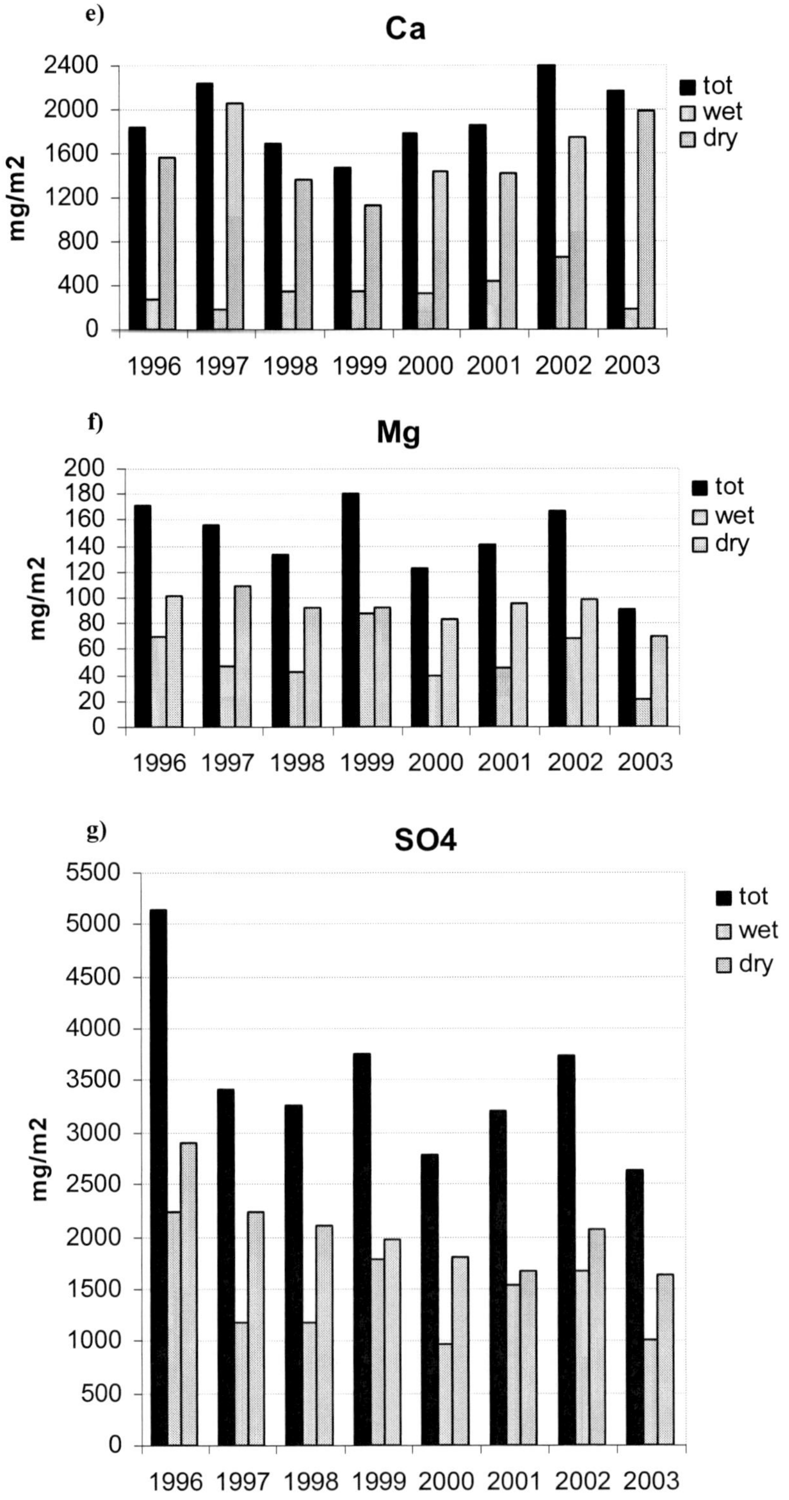

Figure 2. Continued on next page.

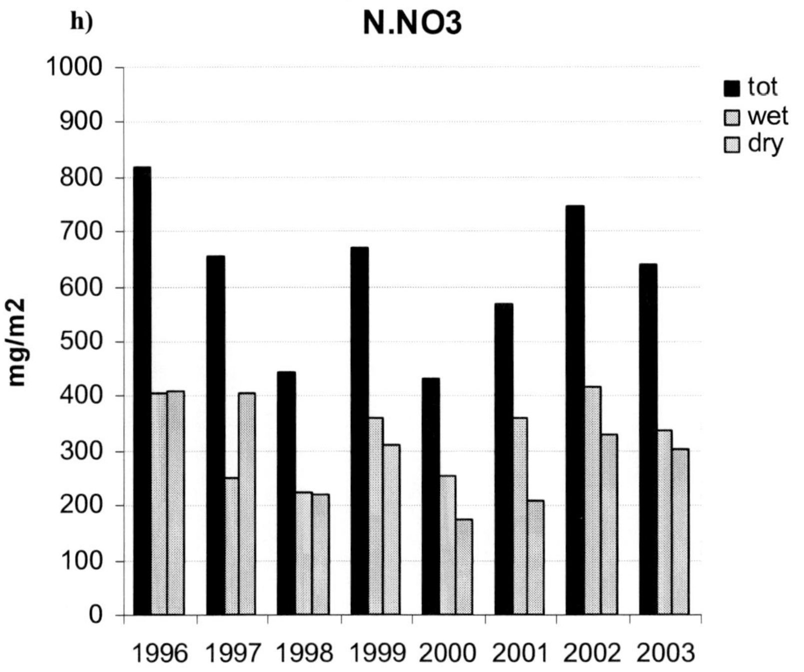

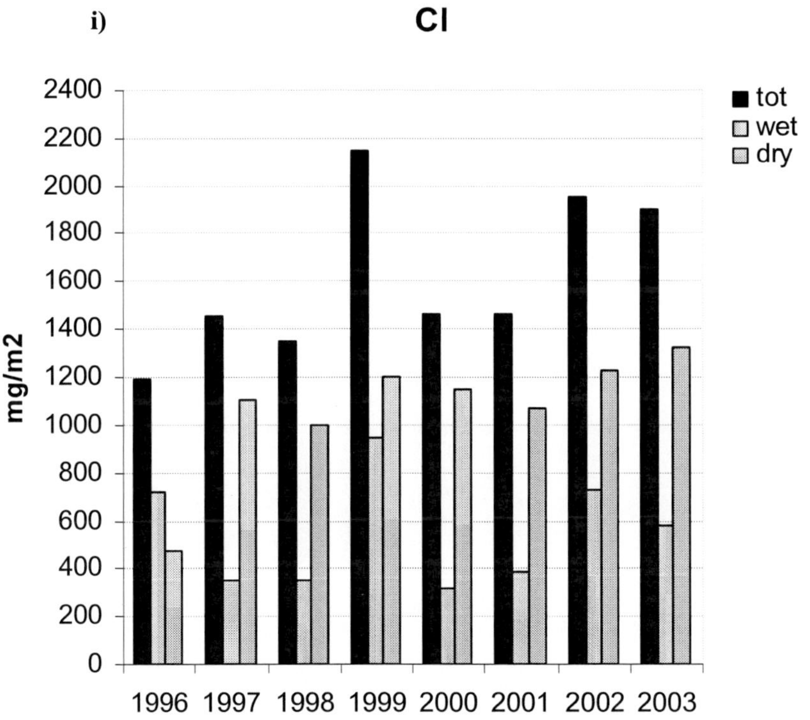

Figure 2. Annual fluxes of total, wet and dry depositions for main ions: a) H^+, b) ammoniacal N, c) Na^+, d) K^+, e) Ca^{2+}, f) Mg^{2+}, g) SO_4^{2-}, h) nitrate N, I) Cl^-. Year 1995 was not considered in the graph because sampling operations started in April.

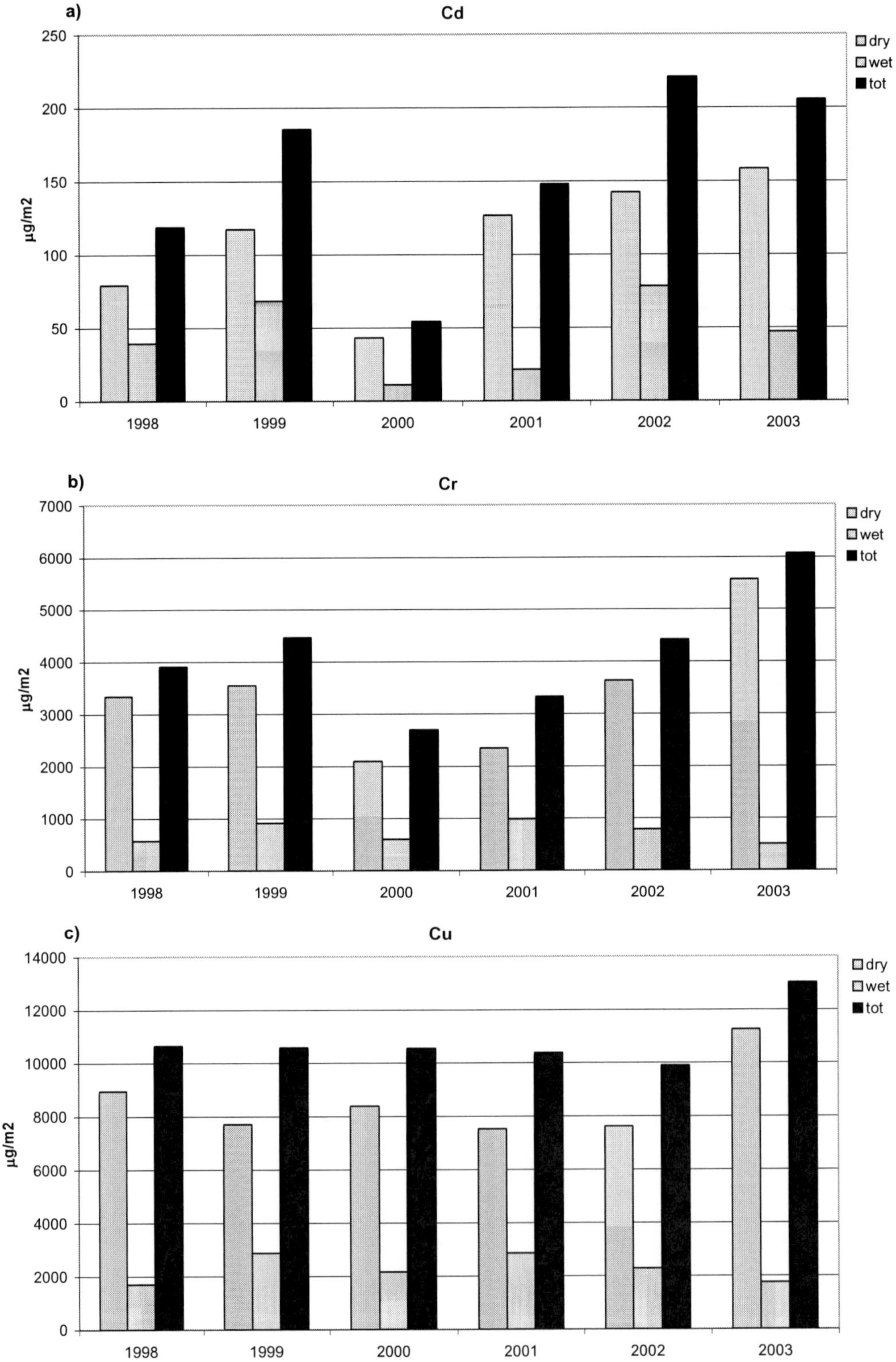

Figure 3. Continued on next page.

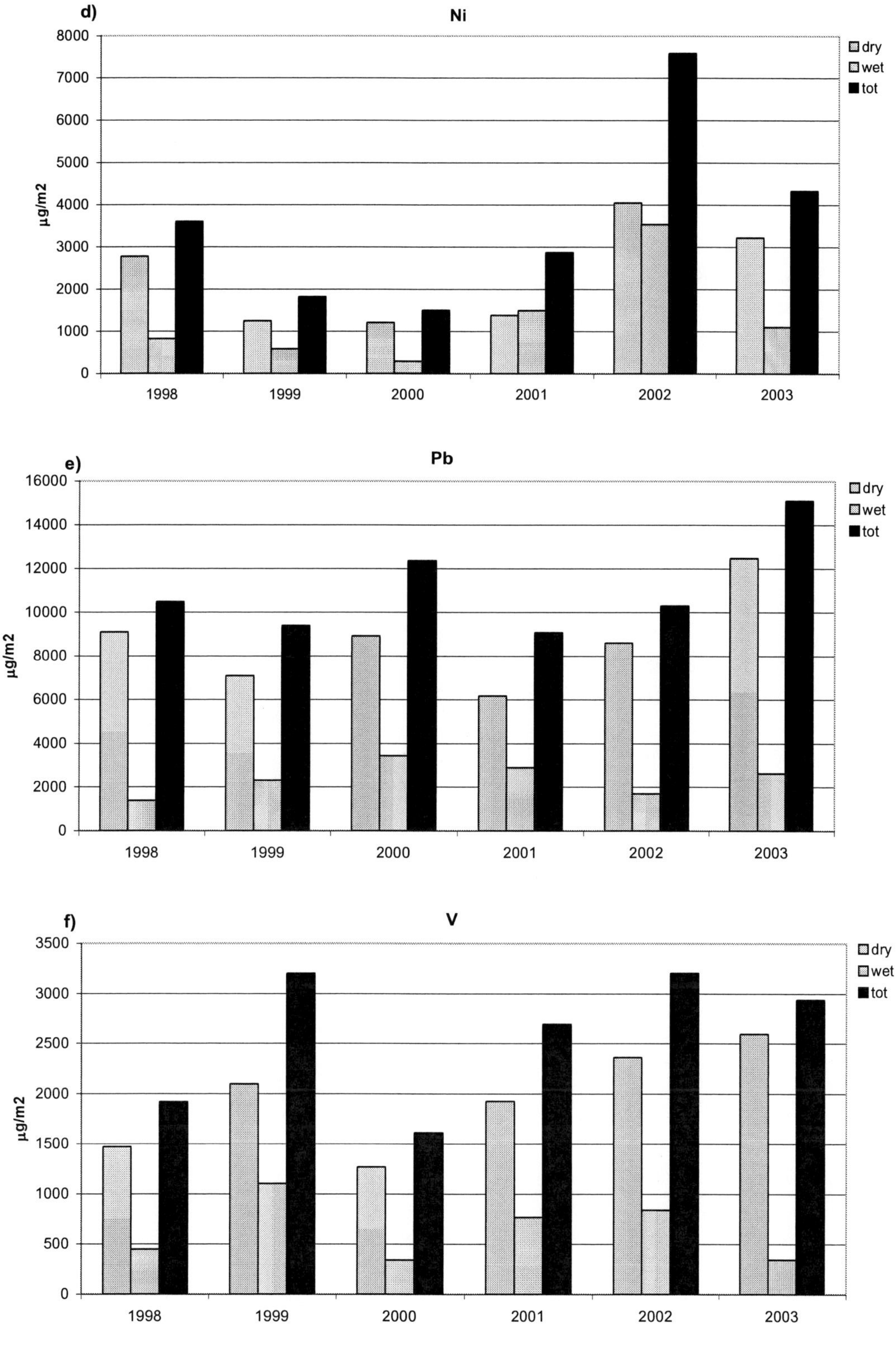

Figure 3. Continued on next page.

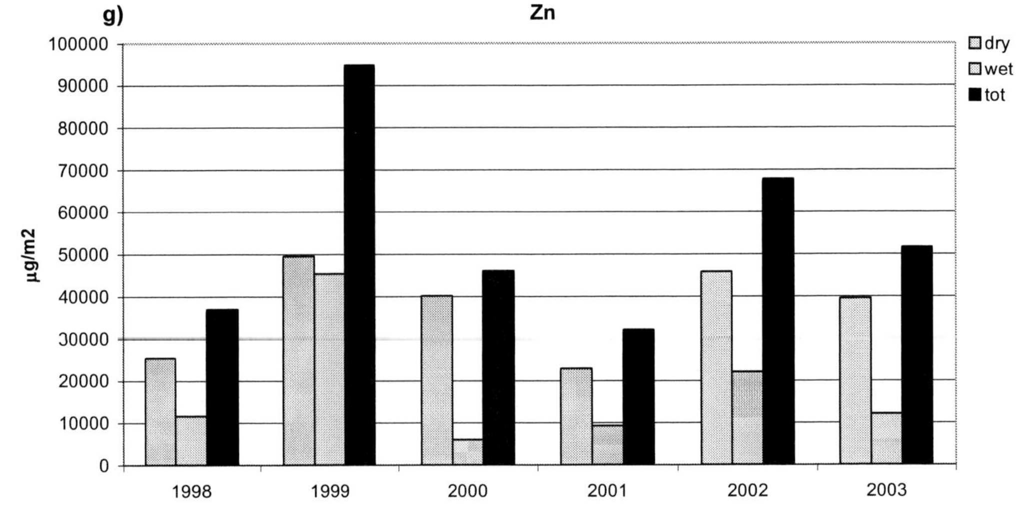

Figure 3. Annual fluxes of total, wet and dry depositions for metals: a) Cd, b) Cr, c) Cu, d) Ni, e) Pb, f) V, g) Zn.

by dry deposition. On average, in the considered period, the dry contribution to the total annual deposition for the main ions is: 81% for Ca^{2+}, 68% for K^+, 65% for Mg^{2+}, 63% for Na^+ and Cl^- and 59% for SO_4^{2-}. The wet contribution prevails only for ammoniacal and nitrate nitrogen (52%) and, above all, for H^+ (95%), that in the dry fraction is, for the most part, neutralized, as shown afterwards. For the analysed metals, the yearly atmospheric removal by dry deposition is always predominant: 73% for Cd, 81% for Cr, 79% for Cu, 78% for Pb, 76% for V, 71% for Zn and 67% for Ni, on average.

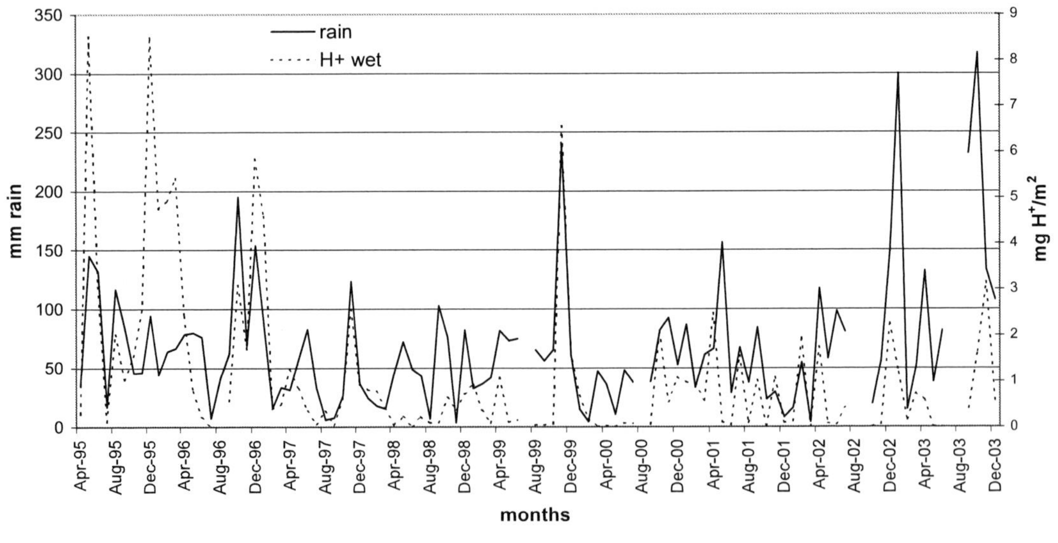

Figure 4. Comparison between monthly rainfall and wet flux of H^+ in the considered period.

Examining in detail each parameter analysed in wet and dry depositions, monthly fluxes show many peaks for H^+ in winter seasons and a marked decreasing trend, in particular from the beginning of our monitoring activity to the first half of 1998. Even though H^+ fallout is almost all due to wet deposition, this trend is not completely explainable with the behaviour of rainfall volumes (Figure 4). It rather seems that rainwater has lowered its acidity due to the clearly decrease, during these years, of sulphate ions (Figure 2.g), both in dry and in wet fractions.

An analogous trend was found for nitrate nitrogen in dry deposition: its fluxes constantly decrease, becoming, in the last four years, lower than the wet ones which, on the contrary, remain nearly constant during the whole survey (29 mg/m^2 per month, on average).

Na^+ shows, on average, a slight increase, due to the dry contribution, whose highest monthly peaks occur after the second half of 1998. However, in general, Na^+ deposition is higher in the spring-summer months.

The monthly trends of wet, dry and total Na^+ deposition are very similar to those of Cl^-; the average Cl^-/Na^+ ratio is equal to 1.8 (g/g), which agrees very well with their ratio in bulk seawater (1.795 g/g) [9], clearly indicating the main common marine origin of these two ions, as confirmed also by the following PCA analysis. Actually, Adriatic Sea is not so far from the monitoring site and, during spring and summer, wind blows prevalently from this direction (Figure 5).

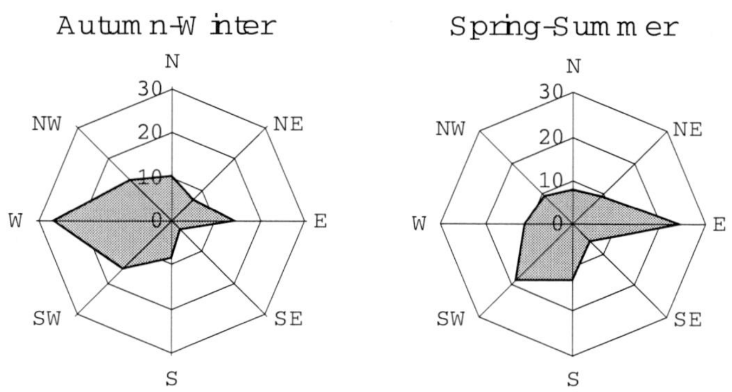

Figure 5. Prevailing wind direction for wind blowing on Bologna (average of the monthly frequencies) - ISTAT Meteorological Statistics.

Mg^{2+}, in the considered period, varies around an average value of 12 mg/m^2 per month, likewise NH_4^+, with an average value of 84 mg/m^2 per month, and K^+ (29 mg/m^2 per month, on average).

Ca^{2+} shows a seasonal trend, with peaks in spring and summer, specifically from April to August, when winds blow more frequently from south-southwest, carrying Saharan dust.

As for the analysed metals, specific trends, neither seasonal or annual, are evident; only Cr shows a clear increase in the considered period.

Acidity and Neutralization in Wet Deposition

Acidity in wet depositions depends on the concentration of both acid forming ions and alkaline species. Therefore, the pH alone gives indication on the free acidity, but cannot exhaustively explain the acid-base phenomena occurring in depositions [32]. Actually, a high pH could be due to the neutralization of the present acidity, rather than absence of acidifying species. If the precipitation acidity is primarily due to sulphuric and nitric acid, the ratio between H^+ and the sum of non-sea salt (nss) sulphate and nitrate should be close to unity [33]. The marine contribution for sulphate was calculated [9] and the average ratio observed for the wet monthly fluxes (0.22), shows that, globally, about 78% of the acidity in wet deposition was neutralized. The $H^+/(\text{nss } SO_4^= + NO_3^-)$ ratio reported in Figure 6 confirms that this neutralization, for wet deposition, is lower in winter.

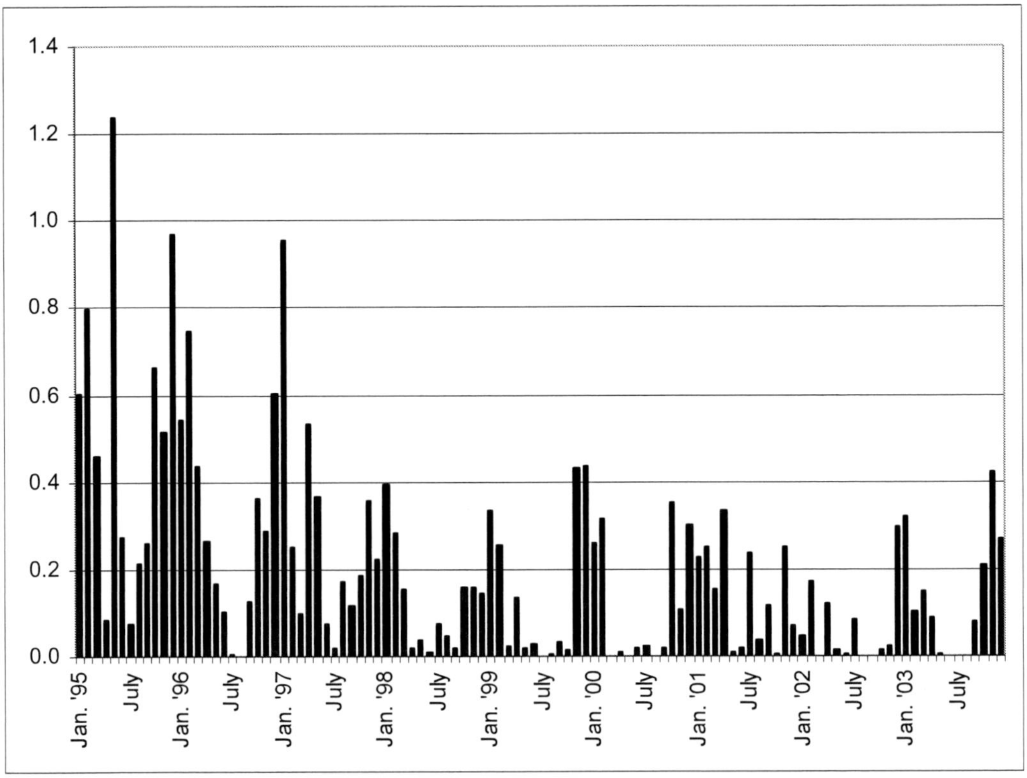

Figure 6. $H^+/(\text{nss } SO_4^= + NO_3^-)$ ratio for wet monthly fluxes (eq/ha).

The average ratio between the equivalent of nss sulphates and nitrates (1.3) shows that the acidifying contribution of these species is near equivalent, while the linear correlation ($r = 0.94$) between ($\text{nss } SO_4^= + NO_3^-$) and ($NH_4^+ + Ca^{2+} + H^+$) indicates that neutralization is mainly due to NH_3 and $CaCO_3$ [32, 33]. Neither NH_3 nor $CaCO_3$ seems to have a seasonal action, as both Ca^{2+} and NH_4^+ show temporal trends analogous to nss $SO_4^=$ and NO_3^-.

Dry depositions are always alkaline; however, considering the acidifying species, sulphates prevail on nitrates ($\text{nss } SO_4^= / NO_3^- = 2.7$).

Metal Solubility

Some interesting remarks can be made observing metal solubility. Since this paragraph on, for a statistically significant evaluation, arsenic and mercury were not considered, because months in which they have been found in a concentration higher than the detection limit in both fractions (soluble and insoluble) were too limited.

The importance of distinguishing soluble and insoluble fractions, both in wet and in dry depositions, is related to the fact that bioavailable fraction (whose interaction with living organisms is greater) is often described as the atmospheric "input" which remains dissolved in a natural water layer [34]; thus, it includes both material dissolved in rain and the dry deposition fraction which dissolves in the aqueous layer maintained as receptor in the sampler.

Table 1. Average of percent monthly soluble fractions for wet, dry and total depositions ($\mu g/m^2$); 1998-2003

	Cd	Cr	Cu	Ni	Pb	V	Zn
Wet	59	26	68	58	41	68	75
Dry	21	3	29	26	8	16	30
Tot	32	8	37	21	36	15	28

The highest contribution of soluble fraction to the total wet deposition is generally greater than that of insoluble one, for all metals but for Cr (26%) and Pb (41%). In particular, Cu, V and Zn show the greatest solubility: respectively, 68%, 68% and 75% (on average) of soluble fraction over the total wet deposition. The order of solubility can be summarised as follows: Zn > V ~ Cu > Ni ~ Cd >> Pb > Cr (Table 1). This result agrees with previous researches [35, 36]. As examples, the graphs of soluble percentages of wet, dry and total depositions for two metals with different solubility are reported: Zn and Cr (Figure 7).

As for dry deposition solubility, it can be easily predicted to be quite different from wet ones, due to the differences in source, composition, structure, interactions with organic and inorganic matrices [37, 38, 39]. First of all, it is evident that in this case, unlike the previous one, insoluble fraction prevails over the soluble one. Comparing, by means of a one-tail t test, the averages of soluble fraction percentages for wet and dry depositions, it can be verified that the latters are significantly lower ($p=0.00005$) for all the analysed metals, confirming that the occurring of rainy events facilitates particle dissolution and the deposition of the most easily bioavailable metals.

However, considering dry monthly fluxes, some exceptions, due to the scarce solubility compared to wet, can be noted in particular for Zn (Figure 7), and, in the first years of monitoring, for Cd, in coincidence with the most rainy months (i.e., November 1999, January 2002 and 2003).

In any case, the order of solubility which is noticed for dry deposition is:

Zn ~ Cu > Ni ~ Cd > V> > Pb > Cr

Zn and Cu prove again to be among the most soluble metals, Pb and Cr among the least.

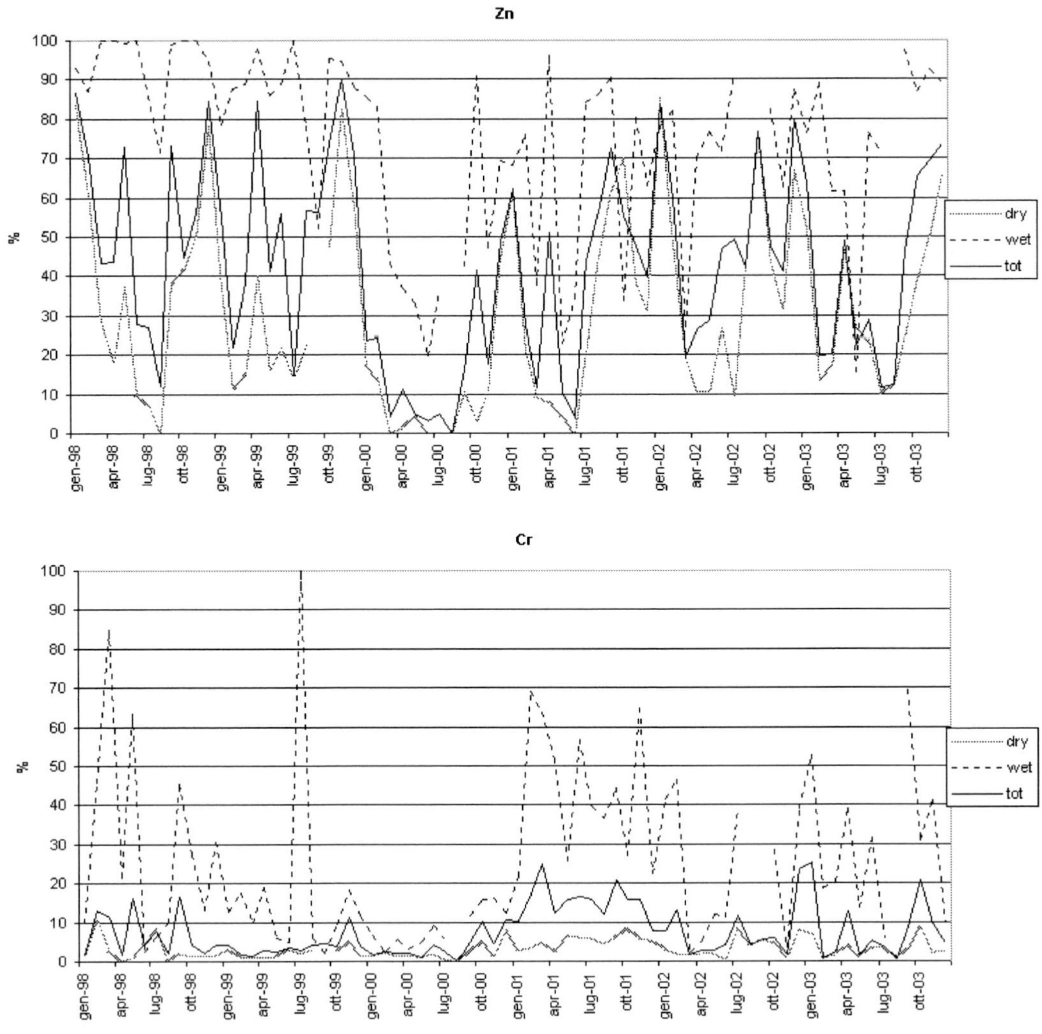

Figure 7. Behaviour of soluble percentages of wet, dry and total deposition for Zn and Cr; 1998-2003.

Correlations between Macroparameters

As for wet depositions, it is reasonable to suppose that there could be significantly positive correlations between the different chemical species and millimetres of rain; actually, Pearson's correlation coefficients ($p<0.001$) are around 0.4 for Ca^{2+}, Na^{+} and Cl^{-}, and about 0.75 for NH_4^{+}, SO_4^{2-}, NO_3^{-}. This suggests that the former ions are mainly associated to coarse particulate, and undergo a gravitational deposition mechanism, scarcely dependent on rain amount, while the latter ones are mostly associated to fine particulate and their removal is substantially due to rainout and washout processes.

Correlation matrices are considered here, for the different deposition fractions; in their calculation outliers have been removed.

In wet deposition as is, almost all species are find well correlated to each other, and it is difficult to make distinctions; this could be due to a masking action given by the rain amount, on which the wet flux of every species depends. So, monthly fluxes have been normalised with respect to mm of rain.

Now, almost all species are still significantly correlated to each other, but with greater differences (Table 2).

Table 2. Correlation matrix – wet deposition normalised over mm of rain; 1995-2003 (*p=0.05*; *p=0.01*; p=0.001)

Wet.nr	H^+	$N.NH_4^+$	Na^+	K^+	Ca^{2+}	Mg^{2+}	$SO_4^=$	$N.NO_3^-$	Cl^-
H^+	1	0.052	*0.210*	***0.286***	***-0.273***	0.054	**0.347**	*0.235*	0.167
$N.NH_4^+$		1	0.132	**0.456**	**0.368**	**0.379**	**0.653**	**0.830**	***0.259***
Na^+			1	*0.206*	-0.033	**0.763**	***0.323***	*0.241*	**0.931**
K^+				1	***0.261***	*0.253*	**0.435**	**0.558**	*0.239*
Ca^{2+}					1	**0.415**	**0.345**	***0.281***	0.002
Mg^{2+}						1	**0.444**	**0.390**	**0.780**
$SO_4^=$							1	**0.693**	**0.318**
$N.NO_3^-$								1	**0.386**
Cl^-									1

In particular, the most significant correlations are found between Na^+/Cl^- (0.93), Mg^{2+}/Na^+ and Mg^{2+}/Cl^- (0.76 and 0.78), thanks to the common marine source. High correlations are also present between the species of prevailing anthropic origin: NO_3^-/NH_4^+ and NO_3^-/SO_4^{2-} (0.83 and 0.69), and NH_4^+/SO_4^{2-} (0.65). The species which is least correlated with the others is Ca^{2+}, in particular with H^+ (-0.27), Na^+, Cl^-. Considering sodium and chloride as sea spray tracers, and H^+ as an indicator of acid deposition, it can be stated that the contribution of these components to the flow of calcium is very limited, and they do not significantly influence its general trend.

As for the species in dry deposition, they are weakly and not significantly anti-correlated with precipitation, except for Ca^{2+} (Pearson coefficient = 0.25).

Table 3. Correlation matrix – dry deposition; 1995-2003 (p=0.05; *p=0.01*; p=0.001)

Dry	H^+	$N.NH_4^+$	Na^+	K^+	Ca^{2+}	Mg^{2+}	$SO_4^=$	$N.NO_3^-$	Cl^-
H^+	1	*0.230*	-0.188	*-0.197*	-0.191	0.001	***0.275***	***0.271***	-0.138
$N.NH_4^+$		1	-0.089	**-0.425**	**-0.430**	-0.030	**0.488**	0.087	0.018
Na^+			1	**0.627**	**0.349**	**0.411**	0.147	-0.172	**0.807**
K^+				1	**0.485**	***0.309***	-0.099	-0.029	**0.564**
Ca^{2+}					1	**0.411**	0.056	0.050	**0.357**
Mg^{2+}						1	**0.433**	0.103	**0.486**
$SO_4^=$							1	**0.424**	*0.211*
$N.NO_3^-$								1	-0.145
Cl^-									1

The highest correlation is found again for Na^+/Cl^- (0.81) (see Table 3), but other significant relations can be observed for K^+ with Na^+, Cl^- and Ca^{2+} (0.63, 0.56 and 0.48), Mg^{2+} with Ca^{2+}, Na^+, Cl^- and SO_4^{2-} (0.41, 0.41, 0.49, 0.43), NH_4^+ with SO_4^{2-} (0.49) and NO_3^- with SO_4^{2-} (0.42). Some significant anti-correlations result in particular for NH_4^+ with K^+ and Ca^{2+} (-0.42 and -0.43). Many of these results agree with those of wet deposition; however, in this case K^+ is apparently better correlated with the species of natural origin.

The correlation matrix of total deposition is affected by both wet and dry contributions, but mostly by the latter, due to the prevailing dry flux over the wet one.

Table 4. Correlation matrix for the ions in total deposition; 1995-2003 (p=0.05; *p=0.01*; p=0.001)

Total	H^+	$N.NH_4^+$	Na^+	K^+	Ca^{2+}	Mg^{2+}	$SO_4^=$	$N.NO_3^-$	Cl^-
H^+	1	**0.437**	***0.286***	-0.055	*-0.228*	**0.321**	**0.697**	**0.502**	**0.339**
$N.NH_4^+$		1	*0.231*	-0.021	*-0.269*	***0.311***	**0.622**	**0.540**	***0.303***
Na^+			1	**0.345**	*0.208*	**0.522**	**0.344**	0.172	**0.859**
K^+				1	**0.509**	*0.231*	0.161	0.193	***0.287***
Ca^{2+}					1	*0.201*	-0.031	0.090	0.113
Mg^{2+}						1	**0.634**	**0.380**	**0.560**
$SO_4^=$							1	**0.658**	**0.390**
$N.NO_3^-$								1	0.169
Cl^-									1

Even in this case (Table 4), the highest correlation can be observed for Na^+/Cl^- (0.86), followed by the ion pairs H^+/SO_4^{2-} and H^+/NO_3^- (0.70, 0.50), NH_4^+/SO_4^{2-} and NH_4^+/NO_3^- (0.62, 0.54), SO_4^{2-}/NO_3^- (0.66), K^+/Ca^{2+} (0.51), Mg^{2+}/SO_4^{2-} (0.63).

As in wet and dry depositions, magnesium is well correlated both with Cl^- and Na^+ (0.56, 0.52 respectively), which is a proof of its considerable marine contribution, and, as in dry deposition, K^+ appears more correlated with species of natural origin.

Principal Component Analysis: Macroparameters

The study of correlations provides useful information to understand the contribution of different sources to the chemical composition of deposition, but an even greater help could come from the application of Principal Component Analysis (PCA). PCA applied on monthly deposition fluxes (1995-2003) has been preceded by the elimination of outliers. Furthermore, as for correlation matrices, wet fluxes have been previously normalised with respect to millimetres of rain.

Varimax rotation has been applied to the principal components explaining more than 10% of the total variance.

As for wet deposition, the first three principal components rotated explain 83% of total variance (Figure 8). The percentage of explained variance of the first two components is similar (36% and 33%): this means that they have a similar influence on the variability of wet deposition.

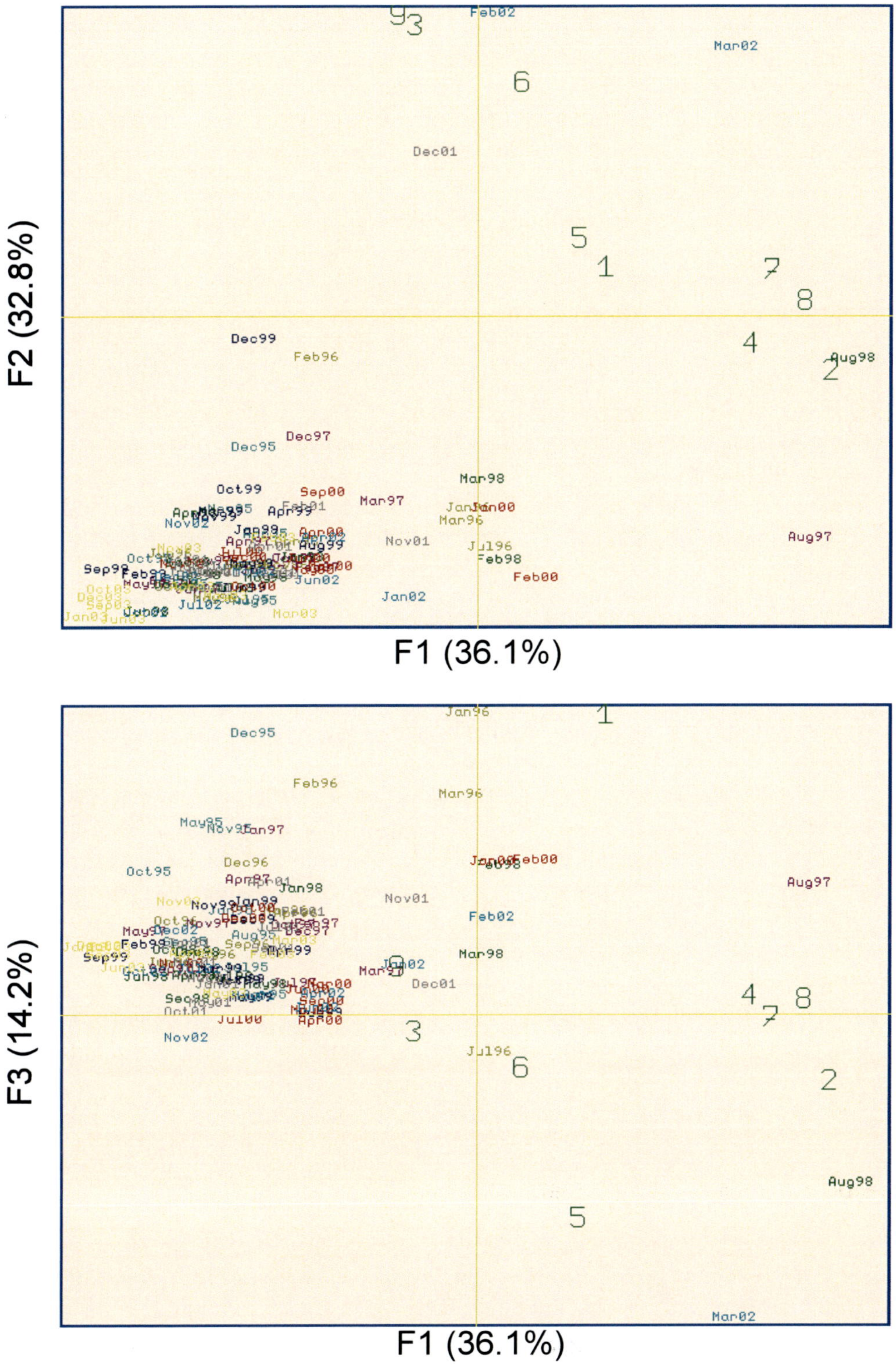

Figure 8. Factor Analysis (Varimax rotation of the first 3 PCs) of wet deposition, normalised over mm of rain.

The first factor is mainly influenced by nitrate, sulphate and ammonium, whose source could be identified in anthropic activities. As for potassium, for which wet deposition contributes only for one third to the total, the relative closeness of the sampling site to the countryside can explain its relationship (already noticed from the correlation matrices) with anthropogenic tracers. Indeed, it can be found associated to nitrates in fertilizers and in manure [40], but also to sulphates in biomass combustion [41]. Since the higher potassium deposition is observed in winter months, the latter hypothesis appears more probable. Another source can be identified in the near municipal incinerator [42], in which a significant percentage of biomass is burnt.

The samples richest in these species are those characterised by the lower rain volumes: 5.9 mm (August 1997), 6.7 mm (August 1998) and 4.2 mm (March 2002); their high concentration in wet deposition could thus be ascribed to the suspended particulate.

The second factor, instead, mainly determined by chloride, sodium and magnesium, represents marine aerosol transportation. Monthly distribution around this factor is not correlated with rain amount.

The third factor, which explains only 14% of variance, is strongly influenced by H^+. It is apparent that months which are affected by the greater acid contribution refer to the first years of sampling, evidencing that acid rain problem has been decreased in the course of time.

Calcium exhibits a clear anti-correlation with H^+, due to the opposite nature of particles or aerosol to which they are associated (basic and generally coarse for Ca^{2+}, acid and fine for H^+).

In the case of dry deposition (Figure 9), the first factor (variance explained 31%) is clearly dominated by marine contribution indicated by the presence of chloride and sodium (it corresponds to the second factor of wet deposition).

Second and third factor have a similar weight, and not much far from that of the first, for dry deposition composition; in fact, they explain both 20% of variance. The second factor is governed by calcium, in the first quadrant, and by ammonium in the forth. Also in this case, calcium could be considered a tracer of natural contribution, mainly given by the fallout of terrigenous aerosol. Ammonium, on the contrary, as species related to the third factor, could be mainly ascribed to anthropic contribution.

Potassium and magnesium give a significant contribution both on the first and on the second factor, suggesting not negligible sources both in marine and in terrigenous aerosol.

While in the PCA obtained for wet deposition, anthropogenic species determine the first factor, here they can be found on the third; a possible explanation could derive from the fact that they are mainly contained in the finest particulate, whose deposition mechanisms do not include gravitational fallout in an appreciable way. Thus, it is reasonable that for the main part these species are removed by wet deposition and their contribution to dry deposition variability is lower than the factors given by natural sources.

More general information can be inferred by the analysis of total deposition (wet + dry), for which the first three principal components explain 75% of variance (Figure 10). Variable distribution resulting after Varimax rotation is hardly predictable from that of the single wet and dry fractions, because it strongly depends on the different partition of each species in the two deposition ways.

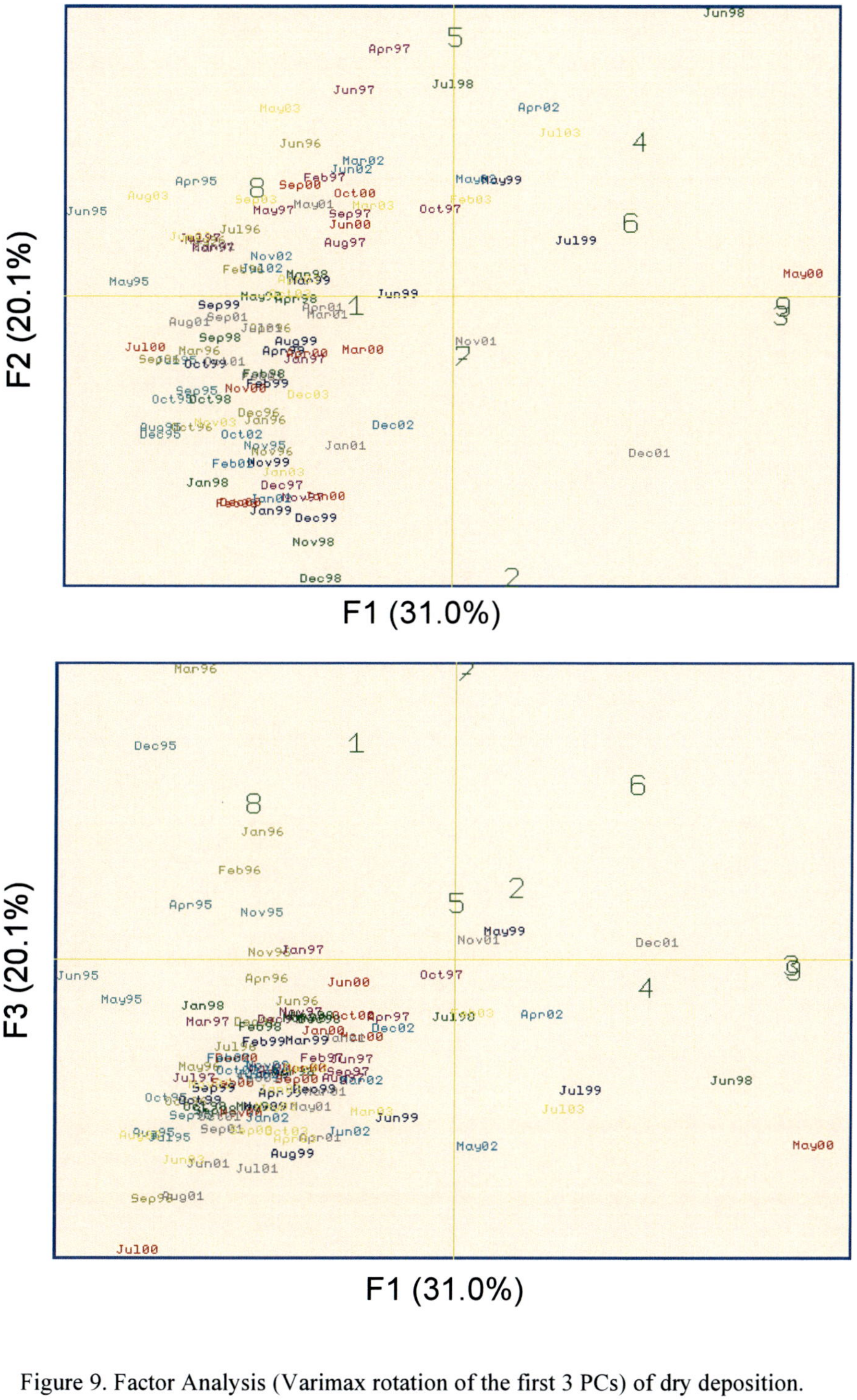

Figure 9. Factor Analysis (Varimax rotation of the first 3 PCs) of dry deposition.

However, the first two factors reflect the same clusters previously observed, with a clear anthropogenic contribution on the first factor (H^+, NH_4^+, SO_4^{2-}, NO_3^-) and marine in the second (in particular Na^+ and Cl^- and, with a lesser extent, Mg^{2+} and K^+). Even in this case, months represented on the first factor (and thus, richest in anthropogenic ions) are almost limited to the first three years of sampling (1995-1997). This agrees with the previous observation of a general decrease of acid species in depositions, and indicates, with regard to acid rain problem, an improvement in the quality of urban atmosphere since the second half of nineties till today.

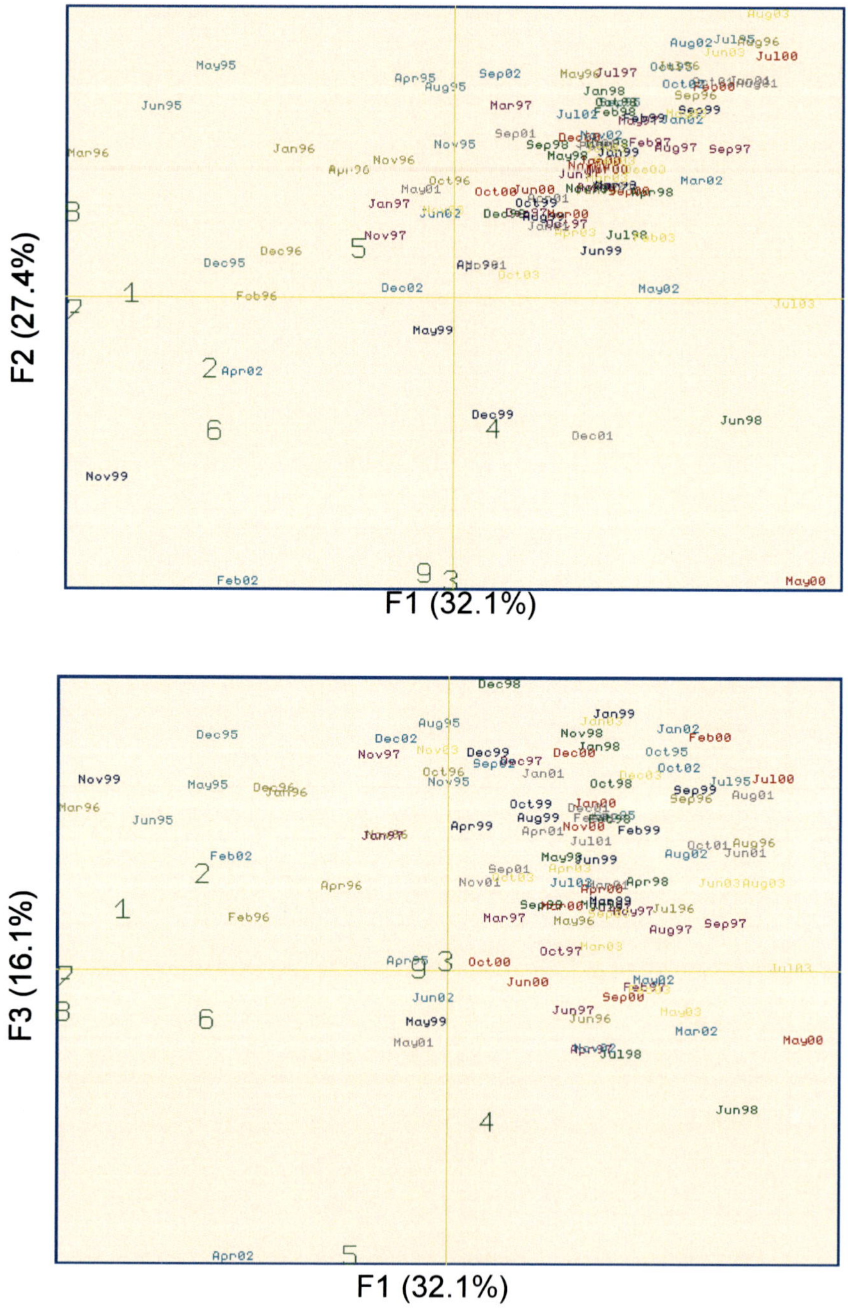

Figure 10. Factor Analysis (Varimax rotation of the first 3 PCs) of total deposition.

The third factor is represented by elements of crustal origin, in particular Ca^{2+} and K^+. Despite this contribution could be simply ascribed to the resuspension of terrigenous particles, a deeper investigation suggest another interpretation for the highest monthly fluxes of calcium. This ion, in fact, is often chosen as tracer of long-range transportation of air masses from Sahara (in particular in South European Countries, in which this phenomenon is quite frequent), sometimes together with Mg^{2+} and K^+ [43, 44, 45]. In many of the months for which high calcium fluxes have been registered, in Northern Italy desert dust transportation events occurred (June '97, June and July '98, September and October '00, May '01, March, April and November '02), some of which have been already studied and reported in literature [46, 47, 48]. Thus, this appears the major contribution recognizable on the third factor; however, high calcium fluxes occurred also in months without known Saharan events, maybe due to soil resuspension on a local scale.

To sum up, major factors affecting chemical composition of wet and dry deposition in the near suburbs of Bologna are three: anthropic, marine and terrigenous.

Anthropic contribution acts on a local scale and is well represented by H^+, nitric and ammoniacal nitrogen and sulphate. Marine aerosol, particularly rich in sodium, chloride and, secondarily, potassium and magnesium ions, could be considered instead a contribution on a regional scale. On the other hand, for the terrigenous factor, characterised by calcium, partly associated to potassium and magnesium, the highest fluxes have been registered during long-range transportation events of Saharan dust.

Considering singly wet and dry deposition it is possible to observe a prevalent removal of anthropic species by the former, due to the presence of these species in the finest particles [32], compared to those deriving by natural sources. Finally, even from the PCA a progressive decrease of acidifying species has been observed.

Correlation between Metals

Correlation matrices for metals in the different deposition ways are here considered; in their calculation outliers were omitted.

All metals in wet deposition (soluble + insoluble), normalised with respect to rain amount, are positively correlated, mostly with a statistical significance (see Table 5).

Table 5. Correlation matrix for heavy metals in wet deposition normalised over mm of rain; 1998-2003 (*p=0.05*; ***p=0.01*; p=0.001)**

Wet.nr	**Cr**	**Cu**	**Ni**	**Pb**	**V**	**Zn**
Cd	**0.473**	0.130	***0.357***	***0.316***	**0.405**	*0.299*
Cr		**0.705**	***0.320***	**0.636**	**0.758**	**0.400**
Cu			0.143	**0.551**	**0.503**	***0.329***
Ni				0.146	***0.329***	0.181
Pb					***0.333***	*0.282*
V						**0.413**

Highest correlations can be found for V/Cr (>0.66), V/Cu (>0.50), Cr/Cu (>0.70), Pb/Cr (>0.58) and Pb/Cu (>0.55) in both soluble and insoluble fractions, and for Cr/Zn (0.68) and Zn/V (0.81) in insoluble one. Even the least soluble metal (Cr) is thus well correlated with the most soluble ones.

In soluble fractions the number of positive and significant correlations is lower; in particular, Cd and Zn results well correlated (0.59), but not with the other metals. Among the others, the most significant correlations occur between Ni e Cr (0.55) and Cu/Pb, Cu/V (>0.45).

Even in the case of dry deposition, metal fluxes in the insoluble fraction are better correlated than in soluble one. In the latter, only Zn/Ni, Pb/Cu and Cu/Cr have a significant correlation coefficient (p = 0.05), while in the former all metals are significantly well correlated (p = 0.001), except for Cd (which shows a significant correlation only with Cr and Ni, p = 0.05). The highest correlations happen among V/Cr (0.77) and Zn/Cr, Pb/Cu, Zn/V, Cu/Cr (all around 0.70).

Being predominant over the soluble one, correlation patterns resulting in the total dry deposition is similar to that found in the insoluble fraction (Table 6). The highest correlations are again V/Cr (0.79) and Zn/Cr (0.70).

Table 6. Correlation matrix for heavy metals in dry deposition; 1998-2003 (*p=0.05*; ***p=0.01*; p=0.001)**

Dry	**Cr**	**Cu**	**Ni**	**Pb**	**V**	**Zn**
Cd	*0.311*	0.202	***0.319***	0.088	0.182	0.072
Cr		**0.635**	**0.504**	**0.578**	**0.789**	**0.698**
Cu			***0.370***	**0.674**	**0.498**	**0.603**
Ni			0.093	0.246	0.209	0.081
Pb				***0.345***	**0.452**	**0.421**
V					**0.402**	**0.456**

In turn, metal similarity in total deposition shows intermediate characteristics between wet and dry, but it is most affected by the latter, due to its quantitative prevalence on the former (Table 7). The highest correlations take place for Cr/V (0.80), Cr/Cu (0.68), Cu/Pb (0.67) and Cu/V, Zn/Cr (>0.52).

Table 7. Correlation matrix for heavy metals in total deposition; 1998-2003 (*p=0.05*; ***p=0.01*; p=0.001)**

Tot.	**Cr**	**Cu**	**Ni**	**Pb**	**V**	**Zn**
Cd	***0.393***	0.237	***0.375***	-0.011	***0.320***	0.178
Cr		**0.680**	***0.335***	**0.468**	**0.799**	**0.539**
Cu			*0.310*	**0.674**	**0.520**	***0.341***
Ni				0.056	***0.334***	**0.425**
Pb					*0.242*	*0.278*
V						**0.461**

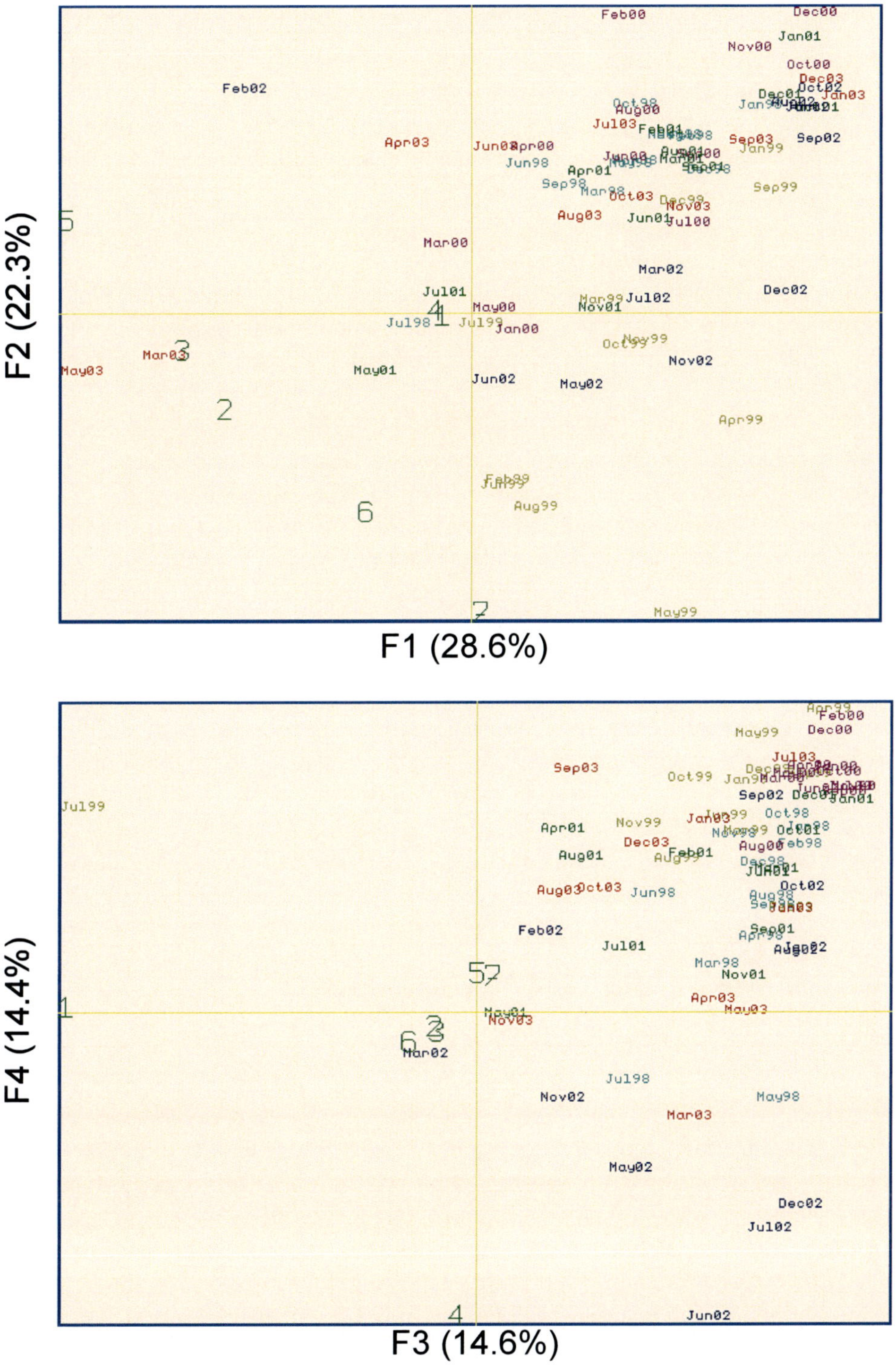

Figure 11. Factor Analysis (Varimax rotation of the first 4 PCs) of total deposition of heavy metals (1998-2003).

To all these analysed fractions of metals, after outlier removal and normalisation, Principal Component Analysis has been applied, followed by Varimax rotation. Wet fluxes have been previously normalised with respect to millimetres of rain.

However, the most interesting considerations can be made observing the behaviour of metals in the total deposition (wet insoluble and soluble + dry insoluble and soluble), which is much similar to the distribution that can be found for dry one (due to the predominance of this way of deposition over the wet one). But the different sources of metals in the environment make the apportionment procedure quite difficult.

As for the total fraction, the first four factors have been rotated, which explain 80% of the total variance (Figure 11). The first factors is determined firstly by Pb and by the couple Cr-Cu. These latter metals prove to be strongly correlated in all insoluble fractions, and also Pb is correlated to those metals in the same fractions. Cu, Cr and Pb could derive from vehicular emissions, as impurities in fuel (Pb in fuels has been permitted till the end of 2001 in Italy), additives in motor oils, brakes or tyres. However, the absence of Zn (which is often taken as vehicular emission tracer) suggests more caution in supporting this hypothesis. Cu, Cr and Pb, elsewhere in literature, are indicated as tracers of the production/use of paints [49], and indeed some factories for varnish production are present around the sampling site. Furthermore, Cr apportion (and sometimes also Cu) is frequently indicated as a crustal contribution [49]; also lead can derive from soil resuspension, possibly distinguishable from the anthropic one by means of an isotopic analysis [50].

The second factor is dominated by V and Zn. They are typically anthropogenic metals: the first derives mainly from fossil fuel combustion, the second from different kinds of vehicular emissions, industrial emissions and from incinerators. Both metals are also employed in metallurgic activities, in particular in steel alloys to improve flexibility or resistance or for material coverage.

The third and the fourth factors are characterised by two "isolated" metals, Cd and Ni respectively.

Correlation Between Micro and Macroparameters - PCA

In order to obtain much information on the different types of metal sources and on their possible counter-ions, a study of correlation matrices and principal components (with Varimax rotation) has been applied simultaneously to micro and macroparameters. Each of these ions, indeed, clearly showed its provenience (anthropic or natural sources).

For uniformity with macroparameters, both correlation and principal components have been determined considering firstly the soluble metal fraction.

In the analysis of principal components the first 3 factors have been rotated, which explain 62% of variance, and have a comparable weight (Figure 12). On the first factor, the anthropic component can be clearly recognised, indicated by the presence of sulphate, nitrate and ammonium; metals which neatly fall in this category are Zn, V and Cd, which, on the other hand, also show good correlation coefficients with the abovementioned species, maybe because they are transported and removed together.

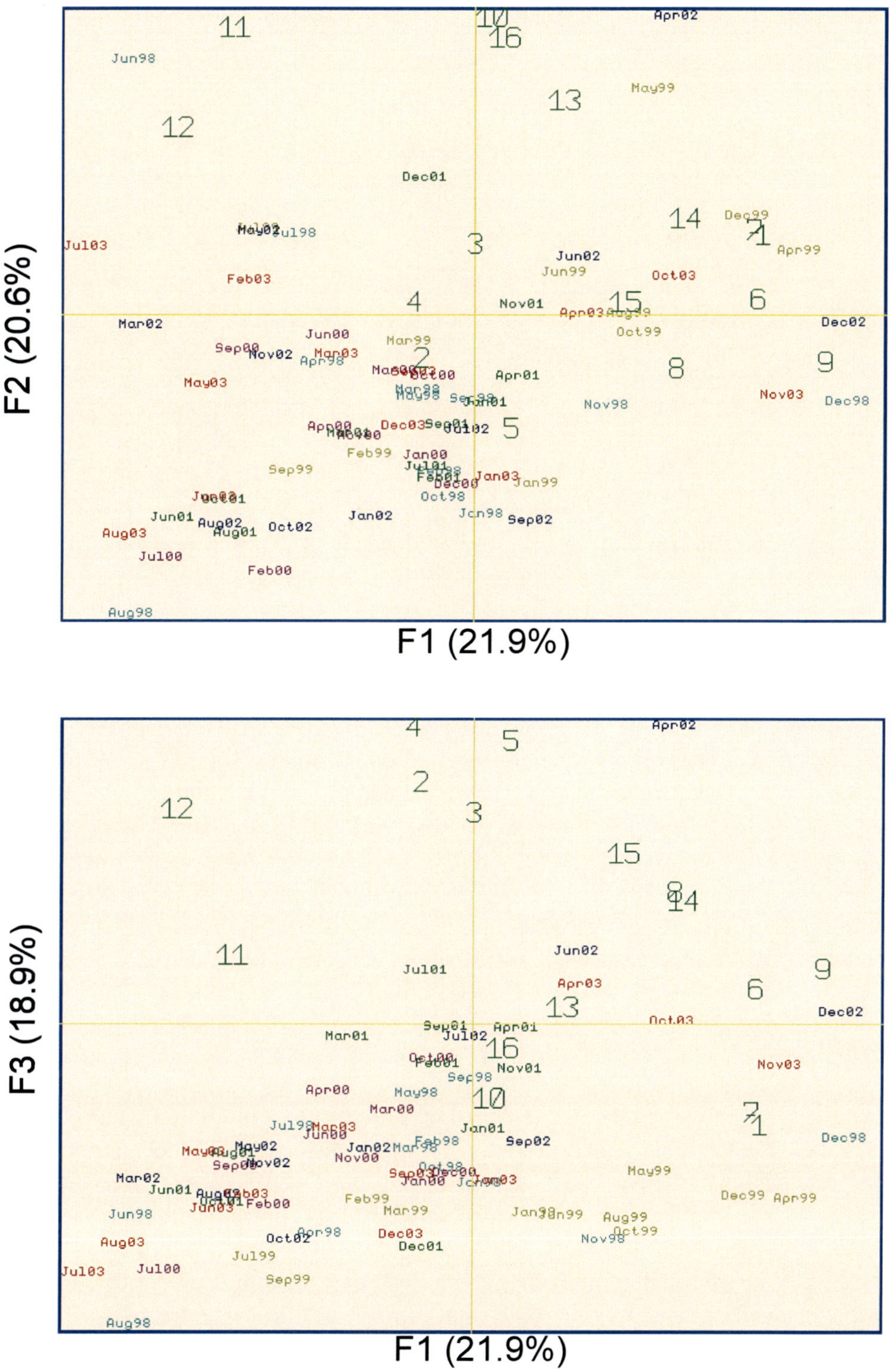

Figure 12. Factor Analysis (Varimax rotation of the first 3 PCs) of the total deposition (soluble fraction for metals) 1998-2003.

Moreover, vanadium is cited in literature as one of the major responsible of primary acidity formation from combustion [51]; correlation coefficient V/H^+ is, indeed, one of the highest (0.52) (Table 8).

Table 8. Correlation matrix – total deposition (soluble fraction for metals); 1998-2003 (*p=0.05*; ***p=0.01*; p=0.001)**

Tot.	H^+	$N.NH_4^+$	Na^+	K^+	Ca^{2+}	Mg^{2+}	$SO_4^=$	$N.NO_3^-$	Cl^-
Cd	0.127	***0.375***	**0.465**	0.139	-0.076	0.188	***0.376***	**0.407**	***0.381***
Cr	***0.358***	0.175	0.151	0.150	0.181	0.148	0.213	***0.357***	0.136
Cu	**0.407**	0.226	**0.460**	*0.272*	0.212	*0.266*	***0.327***	*0.273*	***0.358***
Ni	0.213	***0.327***	-0.109	0.168	0.196	-0.081	*0.289*	***0.368***	-0.086
Pb	**0.459**	**0.434**	0.114	-0.088	-0.023	0.105	*0.298*	*0.310*	0.064
V	**0.525**	**0.612**	*0.293*	0.163	-0.046	**0.485**	**0.661**	**0.522**	***0.343***
Zn	0.074	**0.404**	*0.286*	0.226	-0.046	***0.358***	**0.396**	**0.455**	0.213

The second factor represents sea source, recognisable from the presence of sodium and chloride: no metal derives mainly from that source.

The third factor is chiefly characterised by Ni, Pb, Cr and Cu. The presence of calcium in the opposite quadrant of the anthropic component, suggests that those metals could partly derive from a resuspension of material deposited on the soil and mainly associated to coarse particles. If the total deposition (instead of the lonely soluble fraction) is considered for metals, Cr, Cu, Ni and Pb show a positive and significant correlation with calcium (correlation coefficient around 0.4, $p = 0.001$). It must be reminded, as said before, that also Cr could be a tracer of crustal component and in the vicinity of Bologna, alluvial deposits quite rich in that element are present [52]. To verify if lead is of natural or anthropogenic origin with a significant resuspension contribution, an isotopic analysis could be useful.

Conclusion

Wet and dry depositions in the near suburbs of a city in the Po Valley (Bologna) have been investigated to understand which are main factors influencing their chemical composition. Three principal factors have been identified: anthropic, marine and terrigenous. The first contribution acts on a local scale and can be traced by H^+, nitric and ammoniacal nitrogen, sulphate, and, among metals, mainly Zn, V and Cd.

Generally, dry deposition prevails over the wet ones, i.e., the amount of ions (metals or macroparameters) deposited by dry mechanisms is greater than that due to rainy (or snowy) events. As a consequence, correlation and data dispersion observable from statistical treatments of total deposition results are much similar to those of dry one.

However, the prevailing way of removal of anthropogenic species proved to be the wet one, due to the presence of those species in the finest particulate, compared to those deriving from natural sources. Marine aerosol, particularly rich in sodium and chloride ions, and secondarily potassium and magnesium, can be considered, on the contrary, a regional scale contribution. On the other hand, as for terrigenous factor, characterised by calcium, the main fluxes have been registered during transboundary transportation of Saharan dust.

While specific trends during time cannot be identified for metals, a progressive decrease in free acidity in atmospheric deposition, associated to the reduction in other anthropic species, as sulphates and nitrates, has been observed. This suggests that an improvement in air quality in the urban area occurred, due to recent developments in pollutant emission reduction (in particular, sulphide oxides).

References

[1] Krupa, S.V. *Environmental Pollution* 2002, 120, 565–594

[2] Balestrini, R.; Galli, L.; Tartari, G. *Atmos. Environ.* 2000, 34, 1455-1470

[3] Van Leeuwen, E.P.; Draaijers, G.P.J.; Erisman, J.W. *Atmos. Environ.* 1996, 30, 2495-2511.

[4] Draaijers, G.P.J.; Van Leeuwen, E.P.; de Jong, P.G.H.; Erisman, J.W. Deposition of base-cations in Europe and its role in acid neutralization and forest nutrition. 1996 RIVM Report no. 722108017

[5] Bouwman, A.F.; Van Vuuren, D.P.; Derwent, R.G.; Posch, M. *Water Air Soil Poll.* 2002, 141, 349-382

[6] Morselli, L.; Bernardi, E.; Passarini, F.; Tesini, E. *Ann. Chim.-Rome* 2006, 96, 697-705

[7] Brimblecombe, P. *The effects of air pollution on the built environment*, Imperial College Press, London, UK, 2003

[8] Morselli, L.; Bernardi, E.; Passarini, F. *Environ. Sci. Pollut. R.* 2002, SI 3, 287-288

[9] *Manual on methodologies and criteria for Modelling and Mapping Critical Loads & Levels and Air Pollution Effects, Risks and Trends,* UN/ECE 2004

[10] Nilsson, J.; Grennfelt, P. *Critical loads for sulphur and nitrogen,* Miljørapport 15, Nordic Council of Ministers, Copenhagen, DK, 1988

[11] Erisman, J.W.; Grennfelt, P.; Sutton, M. *Environ. Int.* 2003, 29, 311-325

[12] Hirs, D.; Storvik, G. *Sci. Total Environ.* 2003, 310, 163-170

[13] Morselli, L.; Bernardi, E.; Passarini, F.; Olivieri, P. *J. Phys. IV* 2003, 107, 895-898.

[14] Morselli, L.; Francaviglia, R.; Lepore, L.; Merolli, S.; Passarini, F.; Bernardi, E.; Mezzogori, R. *Ann. Chim.-Rome* 2003, 93, 117-127.

[15] Erisman, J.W.; Mennen, M.G.; Fowler, D.; Flechard, C.R.; Spindler, G.; Grüner, A.; Duyzer, J.H.; Ruigrok, W.; Wyers, G.P. *Environ. Monit. Assess.* 1998, 53, 279-295

[16] Azimi, S.; Ludwig, A.; Thevenot, D.R.; Colin, J.-L. *Sci. Total Environ.* 2003, 308, 247-256.

[17] Morselli, L.; Brusori, B.; Passarini, F.; Bernardi, E.; Francaviglia, R.; Gataleta, L.; Marchionni, M.; Aromolo, R.; Benedetti, A.; Olivieri, P. *Environ. Int.* 2004, 30, 173-181

[18] Vecchi, R.; Marcazzan, G.; Valli, G.; Ceriani, M.; Antoniazzi, C. *Atmos. Environ.* 2004, 38 (27), 4437-4446

[19] Barnaba, F.; Gobbi, G.P. *Atmos. Chem. Phys.* 2004, 4, 2367–2391

[20] Whelpdale, D.M.; Summers, P.W.; Sanhueza, E. *Environ. Monit. Assess.* 1997, 48, 217-247

[21] Morselli, L.; Iannuccilli, A.; Barilli, L.; Olivieri, P.; Francaviglia, R.; Aromolo, R.; Di Carlo, V. *Ann. Chim.-Rome* 1998, 88, 645-655

[22] Mosello, R. *Rapporto sull'attività della Rete RIDEP nel quinquennio* 1988-1992, Documenta No. 44 dell'Istituto Italiano di Idrobiologia del C.N.R., Verbania Pallanza, IT, 1993
[23] Morselli, L.; Mosello, R. *Fresen. Environ. Bull.* 1994, 3 (3), 141-146
[24] Mosello, R.; Baudo, R.; Tartari, G.; Serrini, G.; Muntau, H.; Novo, A. *Metodologie e controlli di qualità per lo studio delle deposizioni atmosferiche in Italia,* Documenta No. 23 dell'Istituto Italiano di Idrobiologia del C.N.R., Verbania Pallanza, IT, 1990
[25] Morselli, L.; Cecchini, M.; Grandi, E.; Iannuccilli, A.; Barilli, L.; Olivieri, P. *Chemosphere* 1999, 38, 899-907
[26] Morselli, L.; Olivieri, P.; Brusori, B.; Passarini, F. *Environ. Pollut.* 2003, 124, 457-469
[27] Bernardi, E. PhD Thesis, Università di Bologna, 2005
[28] Zobrist, J.; Wersin, P.; Jaques, C.; Sigg, L.; Stumm, W. *Water Air Soil Poll.* 1993, 71, 111-130
[29] Morselli, L.; Brusori, B.; Cecchini, M.; Olivieri, P.; Silingardi, D.; Passarini, F. *Ann. Chim.-Rome* 2001, 91, 459-469
[30] Polkowska, Ż.; Astel, A.; Grynkiewicz, M.; Górecki, T.; Namieśnik, J. *J. Atmos. Chem.* 2002, 41, 239-264
[31] Forina, M.; Lanteri, S.; Armanino, C. *Q-PARVUS Release* 3.0, *an extendable package of programs for explorative data analysis, classification and regression analysis*, Dip. Chimica e Tecnologie Farmaceutiche, University of Genova, IT, 2001. Available on-line (free, with manual and examples) at http://parvus.unige.it.
[32] Al-Momani, I.F.; Tuncel, S.; Eler, U.; Ortel, E.; Sirin, G.; Tuncel, G. *Sci. Total Environ.* 1995, 164, 75-85
[33] Kaya, G.; Tuncel, G. *Atmos. Environ.* 1997, 31, 3985-3998.
[34] Sandroni, V.; Migon, C. *Chemosphere*, 2002, 47, 753-764.
[35] Golomb, D.; Ryan, D.; Eby, N.; Underhill, J.; Zemba, S. *Atmos. Environ.* 1997, 31, 1349-1359
[36] Guieu, C.; Chester, R.; Nimmo, M.; Martin, J.M.; Guerzoni, S.; Nicolas, E.; Mateu, J.; Keyse, S. *Deep-Sea Res Pt II*, 1997, 44, 655-674
[37] Chester, R.; Nimmo, M.; Fones, G.R.; Keyse, S.; Zhang, J. *Atmos. Environ.* 2000, 34, 3875-3887.
[38] Hoffmann, P.; Dedik, A.N.; Deutsch, F.; Sinner, T.; Weber, S.; Eichler, R.; Sterkel, S.; Sastri, C.S.; Ortner, H.M. *Atmos. Environ.* 1997, 31, 2777-2785.
[39] Spokes, L.J.; Campos, M.L.A.M.; Jickells, T.D. *Atmos. Environ.* 1996, 30, 3959-3966
[40] Tamminga, S. *Lives. Prod. Sci.*, 2003, 84, 101-111
[41] Calloway, C.P.; Li, S.; Buchanan, J.W.; Stevens, R.K. *Atmos. Environ.* 1989, 23, 67-69.
[42] Dzubay, T.G.; Stevens, R.K.; Gordon, G.E.; Olmez, I.; Sheffield, A.E.; Courtney, W.J. *Environ. Sci. Technol.*, 1988, 22, 46-52.
[43] Schwikowski, M.; Seibert, P.; Baltensperger, U.; Gäggeler, H.W. *Atmos. Environ.* 1995, 29, 1829-1842.
[44] Guerzoni, S.; Molinaroli, E.; Chester, R. *Deep-Sea Res Pt II*, 1997, 44, 631-654.
[45] Alastuey, A.; Querol, X.; Chaves, A.; Ruiz, C.R.; Carratala, A.; Lopez-Soler, A. *Environ. Pollut.*, 1999, 106, 359-367
[46] Bonasoni, P.; Van Dingenen, R.; Calzolari, F.; Cristofanelli, P.; Putaud, J.P.; Bonafè, U.; Evangelisti, E.; Balkanski, Y. Proceedings, "A Changing Atmosphere. 8th European

Symposium On the Physico-Chemical Behaviour of Atmospheric Pollutants", Turin (Italy), 17-20 September 2001.
[47] Balkanski, Y.; Bauer, S.E.; Van Dingenen, R.; Bonasoni, P.; Schulz, M.; Fischer, H.; Gobbi, G.P.; Hanke, M.; Hauglustaine, D.; Putaud, J.P.; Stohl, A.; Raes, F. *Atmos. Chem. Phys. Discuss.*, 2003, 3, 1753-1776.
[48] Barnaba, F.; De Tomasi, F.; Gobbi, G.P.; Perrone, M.R.; Tafuro, A. *Atmos. Res.*, 2004, 70, 229-259.
[49] Pekey, H.; Karaka, D.; Bakoglu, M. *Mar. Pollut. Bull.*, 2004, 49, 809-818.
[50] Wong, C.S.C.; Li, X.D.; Zhang, G.; Qi, S.H.; Peng, X.Z. *Atmos. Environ.* 2003, 37, 767-776.
[51] Primerano, P.; Campisi, I.; Di Pasquale, S.; Corigliano, F. *Atmos. Environ.*, 1999, 33, 3551-3558.
[52] Dondi, M.; Ercolani, G.; Guarini, G. *Mineralogica Petrographica Acta*, 1993, *XXXVI*, 129-142.

In: Acid Rain Research Focus
Editor: Lyman G. Roglesfield, pp. 105-124
ISBN: 978-1-60456-373-3

Chapter 4

Bioremediation of Soil and Water Acidified by Acid Rain with Fungi Cells

Naofumi Shiomi[1] and Shigeo Katoh[2]
[1]School of Human Sciences, Kobe College
[2]Graduate School of Engineering, Kobe University

Abstract

Acid rain damages many kinds of plants and living things by reaching cations from soil and lake water, and has become a serious problem in the world. To protect soil and lake water from acidification by acidic rain, calcium carbonate powder has been used for neutralization of acid in Europe and Canada, but they cause other problems. Thus, we reviewed novel bioremediation methods of acidic soil and water with microbes in this report. We screened microorganisms showing a high ability to neutralize acids. These strains of fungi that were isolated from soil microorganisms showed high resistance to acid and a high ability of neutralization. The fungi could grow in acidic media (pH 4.0) containing several kinds of acids and increased pH up to around 8.0. When they grew in LB medium (pH4.0) adjusted with nitric acid, NH_4^+ in the medium increased and NO_3^- decreased. Therefore, it is considered that the neutralization of nitric acid was caused by both a secretion of NH_4^+ and by the assimilation of NO_3^-.

In addition, we proposed a self-immobilization method (BSIS), in which a *Bacillus* strain forming aggregates and a strain bioremediating pollutants were co-immobilized in a shallow layer in soil. Bioremediation in the shallow layer would be effective for protection of ecosystems in soil. By the BSIS method, fungus cells could be co-immobilized with *Bacillus subtilis* in a shallow layer of model soil. The immobilized cells could successfully neutralize acidic solutions flowing through packed soil. This means that BSIS is applicable to neutralization of acid rain caused by SOx and NOx for a relatively long time. We also applied a fungus *Rhizopus* strain to bioremediation of acidified water. The strain could grow and neutralize acid in pond water. To neutralize a large amount of water in a short period, the strain was immobilized in a cellulose tube. Although the volume ratio of the cell suspension to pond water was only one-fortieth, the pH value increased from 4.0 to 7.8 within 1 hour, and nitrates decreased below 0.02 g/l. These results suggest that the immobilized *R. delemar* cells could continuously neutralize pond water and this immobilization method is useful in neutralizing acidified water for a long period.

1. Acidification of Water and Soil by Acid Rain

Serious problems of acidification of water in ecosystems have widely been reported, and fish are most sensitive to acidified water. Individuals with bended backbone and unusual skeleton were found in acidified lakes, because magnesium and calcium ions required to form their skeleton were leached out. The report on acidified Little Rock Lake showed that many species of fish living there were affected below pH 5.0 (Swenson *et al.* 1989). According to the assessment of the status of fish population in 3821 lakes in Norway, Sweden and Finland, many species of fish were already lost by environmental changes caused by acidification (Tammi *et al.* 2003). At extremely acidic lake (pH 3.5) in Japan, no fish can survive except only one special fish, Osorezan dace, which could adapt to low pH (Hirata *et al.* 2003).

Soil is also gradually acidified by acid rain. Soil particles are negatively charged and can adsorb cations (Ca^{2+}, Mg^{2+}, K^{+} and Al^{3+}), in case that pH value of soil is kept close to neutral by the buffering action of carbonate and soil microorganisms. However, excess supply of acid from acid rain makes the pH value of soil acidic. Plants and forests in Europe and northern America have been damaged by acid rain (Matzner *et al.* 1983, Evans 1982). Several factors that damage plants are reported. One factor is a direct effect of acids on leaves of plants. Acids melt a surface layer of leaves and penetrate into them. As a result, the injured leaves die or tend to be pathogenic. Another factor is an indirect effect on soil. Acidification of soil causes loss of basic nutrients and minerals, such as Mg^{2+} and K^{+}, which were important for the growth of plants, and accumulation of Al^{3+} which were harmful to plants. In addition, the acidification caused the death or decrease in root nodule bacteria that help supply of nutrients to roots and small animals producing nutrients in soil. Moreover, important nutrients and minerals such as Ca^{2+}, Mg^{2+} and K^{+} are supplied by microbial degradation of sediments of withered plants on soil surface (so-called Ao zone), but the supply of them also suppressed because of death of soil microorganisms on soil surface. Consequently, shortage of nutrients and minerals in soil cause damages on plants. The direct and indirect factors synergistically damage forests and plants.

The challenge of decrease of acids emission has been practiced since late 1970s. Emission regulation of fossil fuels was effective for decreasing acid pollution. The Canada-U.S. Air Quality Agreement was agreed to reduce SO_2 emission in North America, and both countries have made efforts to reduce the emission (Jeffries *et al.* 2003). Consequently, slow recoveries from acidic conditions in some lakes were observed (Warby *et al.* 2005). For example, in lakes at Killarney Park, which were highly acidic in 1970s, noticeable increase in pH values to above pH 6.0 and decrease in SO_4^{2-} concentrations were reported because of a reduction of air pollution in Sudbury (Keller *et al.* 2003, Larssen *et al.* 2003, Snucins *et al.* 2001). Decreases in acidification by emission control in Europe were also reported. Finnish lakes (Forsius *et al.* 2003), lakes in Bohemian Forests (Vrba *et al.* 2003), Sudetic lakes (Schindler 1988, Sienkiewicz *et al.* 2006) and some lakes in England (Tipping *et al.* 2002) were highly acidic by air pollution in 1970s, but some of the lakes have been slowly recovered from acidic conditions since the degree of air pollution was decreased in the 1980s. Acid neutralizing capacities in lake water were also estimated by monitoring (Berg *et al.* 2005) and by applying dynamic acidification model such as MAGIC3 (Larssen *et al.* 2003, Sverdrup *et al.* 2005). In spite of these efforts, most of lakes in the North America and Europe were still highly acidic because of chronic and episodic acidification (Warby *et al.* 2005).

Neutralization with basic chemicals is one of the general methods for remediation of such acidified soil and water. Calcium carbonate or calcium hydroxide powder, which had been used to maintain neutral pH values of soil in plantations of crops, was applied to remediation of acidified soil and water (Stevens and Dunn 2004, Materechera and Mkhabela 2002, Seoane and Leiros 2001). Many lakes in Europe were neutralized by spraying them from a small ship or helicopter to lakes to protect fish and others. However, this method has some problems. Acidified water constantly flows into lakes and fall on soil. Consequently, calcium carbonate powder spread over lakes or lands could not maintain pH values at neutral for a long period, and thus this method needed high labor and material costs. In Sweden, about three million dollars were spent for remediation of lakes with calcium carbonate powder. In Japan soil of Ashio Mountain, where there were many refineries of copper until 1973, became acidic condition, and trees in this mountain area were killed by SO_2 exhausted from refineries. Spraying $CaCO_3$ and nutrients in a project was tried, but they were lost after rainfall without expected effects. Although more than ten billion yen were invested to the project, the pH values of soil were still low. In addition, an excess supply of calcium ion inhibits absorption of magnesium ion in plants, and thus supply of magnesium ion is also required. Therefore, it is necessary to develop effective bioremediation methods for neutralization of acidified soil and water with low cost and for a long period. In this article, bioremediation of acidified soil and water with fungi were reviewed.

2. Acid Resistant Microorganisms

Values of pH in soil and water are usually maintained at around pH 7 by metabolisms of microorganisms, algae and planktons. For example, the pH values of the Akashi River and some pond in Kobe Prefecture in Japan are kept at pH 6-8. However, these living systems are often disturbed by excess amounts of acids in Europe, and the pH values in soil and water have been gradually decreased. Thus, it is important to clarify the mechanisms of neutralization with these microorganisms and to strengthen their abilities.

Many microorganisms resistant to acids are well known (Cotter and Hill 2003). Some archaebacteria, such as *Thermoplasma acidophilium*, *Thiobacillus sp.* and *Acidiphilium sp.*, living in acidic hot spring show high resistance to extremely low pH (pH 1-3), because their cell membrane is composed with special molecules (Searcy 1976, Dalrand e*t al.* 1970 , Elferrink e*t al.* 1992) . Acidic conditions are most desirable for these strains, and these strains cannot survive above pH4-5. *Lactobacillus casei* and *Acetobactor acet*i, which produce lactic and acetic acids respectively, are well known as strains for the production of yoghurt and vinegar. They also showed resistance to low pH and decrease pH value of medium during cultivations. Judging from these facts, these kind of strains can show acid resistance but do not increase environmental pH, because they have mechanisms to endure and grow under acidic conditions.

On the other hand, some reports characterized yeast strains that show high resistance to acids and, at the same time, increase their environmental pH values during cultivation. For example, *Rhodotorula glutinis* R-1 showed resistance to acid and also showed the ability of neutralizing acidic medium, although it was not high (Nguyen *et al.* 2001). Thus, we newly screened and characterized such microorganisms (Shiomi *et al.* 2004).

For screening of microorganisms showing neutralization ability of acids, cells in soil samples collected from 80 sites in Japan were spread on LB (1% bacto peptone, 0.5% yeast extract, 0.5% NaCl, 0.1% glucose and 0.01% of methyl red) agar plates adjusted at pH 4.0 with nitric acid. A strain showing a neutralization characteristic can be screened by examining color of halo, because the color of methyl red changes from red to yellow when a pH value becomes above approximately 5.0. After incubation at 30 °C for 3 days, three strains NPH1, NPH2 and NPH6 were isolated, which could change the color of the plate to yellow over a large area probably by secretion of basic materials. From the morphological analysis and DNA sequence of the 28S rRNA, the isolated three strains NPH1, NPH2 and NPH6 were identified as filamentous fungi, *Mucor sp.*, *Aspergillus fumigatus* and *Aureobasidium pullulans,* respectively. Figure 1 shows a photograph showing typical change in color of the screening plate, and microscopic photograph of strains NPH2 and NPH6. To clarify growth characteristics of these strains under acidic conditions, the three strains were cultured in the LB medium at pH 2.5 and 3.0. The strains NPH2 and NPH6 could grow slowly at pH 3.0 but could not grow at pH 2.5. Strain NPH1, which showed high resistance to low pH, could grow slowly even at pH 2.5 as shown in Figure 2. The pH of the medium finally reached 8.0 after 14 days.

Microorganisms that show a high ability of neutralization were also screened from thirty strains obtained from several organizations (Nabeshima *et al.*). Figure 3 shows the typical results for six strains, when these strains were cultured in an LB medium adjusted to low pH (pH 3.0) with nitric acid. *Pseudomonas putida* and *Sacchromyces cerevisiae* could not increase the pH value of the medium, but yeast *Yarrowia lipolitica* and fungi strains, such as *Rhizopus delemar*, *Aspergilus oryze* and *Mucor neimalis*, showed a high ability to neutralize acid. Other fungi strains such as *A. niger*, *A sojae*, *A. javanicus*, *R. chinesis, M. gheimalis, M. javanicus, P. chysagehum* and *P. sitrihum* also showed the neutralization ability. These results suggested that fungi and yeast *Yarrowia* (or *Candida*) strains generally showed resistance to low pH value and high neutralization ability, although some bacteria isolated from soil (Yasuda. *et al.* 2004) showed the high neutralization ability. A low pH value is undesirable condition for yeast and fungi because optimal pH values of these strains are around pH 7, and thus they had to neutralize acids for keeping suitable growth conditions.

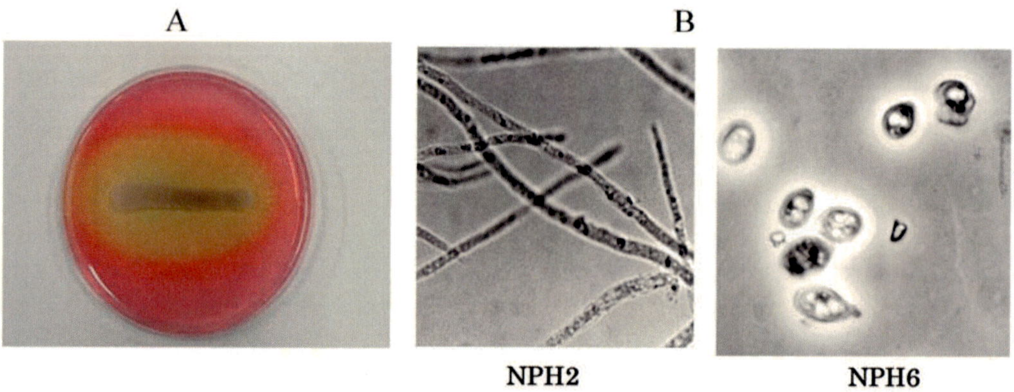

Figure 1. (A) NPH6 cells grown on an agar plate with methyl red. Medium: LB medium (pH 4.0) containing nitric acid and methyl red. (B) Micrographs of isolated strains NPH2 and NPH6.

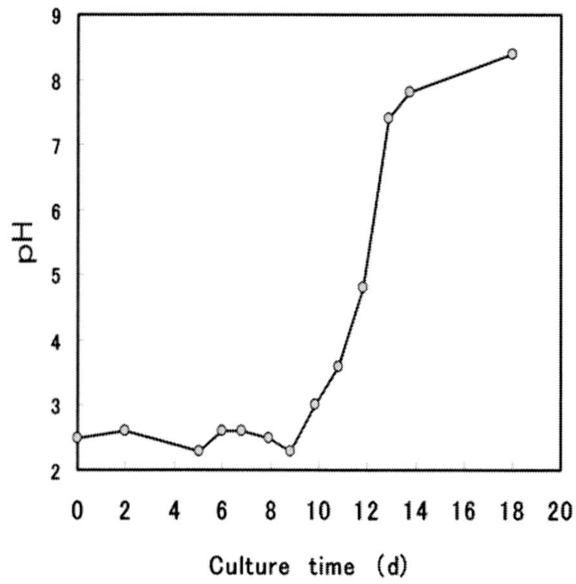

Figure 2. Time course of the change in the pH of LB medium used for growing strain NPH1 (initial pH 2.5, adjusted using nitric acid). Initial cell density; 6 g-dry cells/l.

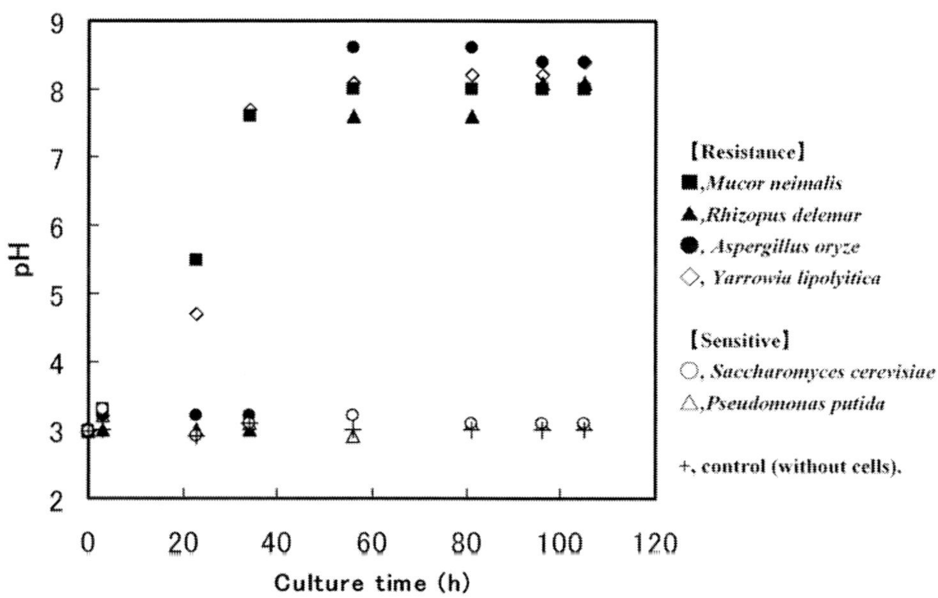

Figure 3. Change in pH with various microorganisms grown in LB medium adjusted at pH 3.0 with nitric acid. Symbols: ■, *Mucor neimalis*; ▲, *Rhizopus delemar*, ●, *Aspergillus oryze*; □, *Yarrowia lipolytica*; ○, *Saccharomyces cerevisiae*; □, *Pseudomonas putida*; +, control (without cells).

3. Mechanism of Neutralization of Acids by Fungi

To clarify the mechanisms of neutralization of fungi, the characteristics of fungi were studied by use of *Rhizopus delemar* NBRC4697. Figure 4 shows the pH change when *R. delemar* NBRC4697 cells were cultured in 50ml of LB media at pH 4.0 adjusted with nitric, sulfuric and hydrochloric acids. *R. delemar* could increase the pH vales adjusted with sulfuric and hydrochloric acid, as well as nitric acid, and therefore the pH increase seemed to be caused by the secretion of basic materials. When 12 g of wet *Rhizopus delemar* cells were incubated in 50 ml of distilled water for 77 h, basic material were secreted or leaked from cells. The pH value of the supernatant increased from 5.1 to 8.2, and about 50 ml of 0.01 N hydrochloric acid was needed to decrease to the initial pH value. The NH_4^+ concentration in the supernatant was 0.04 mol/l of NH_4^+, which means that around 200 ml of hydrochloric acid would be necessary for the neutralization of the supernatant. The results showed that ammonia was an important basic material in cells, but many types of basic and acidic materials were contained and showed buffering action in the supernatant.

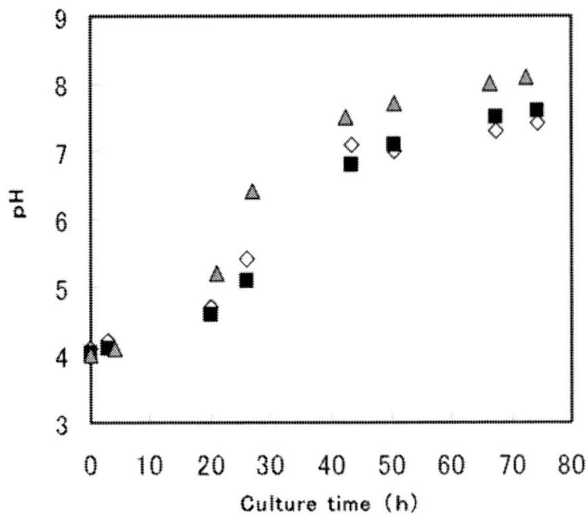

Figure 4. Acid neutralization of *R .delemar* NBRC4697 in LB media adjusted to pH 4.0 with nitric (□), sulfuric (□) and hydrochloric (□)acid.

The increasing pH rate in Figure 4 was higher in nitric acid than in sulfuric and hydrochloric acids. To confirm the mechanism of increase in the pH, *R. delemar* NBRC4697 was grown in LB medium (pH 4.0) that had been adjusted nitric acid. As shown in Figure 5, *R. delemar* NBRC4697 not only secreted NH_4^+ to the medium but also assimilated nitrate (NO_3^-). When *R. delemar* NBRC4697 cells were cultured in a Davis-Miligano minimal medium containing nitric acid or ammonia (pH 7.2), *R. delemar* NBRC4697 cells could grow by nitric acid or ammonia as the sole nitrogen source. We also studied on the mechanisms of neutralization of the other fungi NPH2 and NPH6. These stains could secrete basic materials for neutralization, grow in the medium at pH 4.0 and increase the pH of the medium to approximately 8.0. These strains could also grow by utilizing nitric acids as a sole nitrogen sources. Consequently, the mechanisms of neutralization in NPH2 and NPH6 were similar to those of *R. delemar* NBRC4697. The results suggested that the increase in pH of the culture broth by fungi stains was caused by both a secretion of ammonia and by the assimilation of NO_3^- in the culture broth. It is reported that fungus *F. oxysporium* has the ability of denitrification and changes NO_3^- to N_2 under anaerobic condition (Takaya 2002, Zhou *et al.* 2002). It also has the reduction ability of NO_3^- to NH_4^+ coupled with catabolic oxidation of electron donors under anaerobic condition. Therefore, nitrate taken into *R. delemar* NBRC4697 cells might be changed to ammonia by denitrification. A part of the resulting ammonia may be used for the production of aspartic acid, glutamic acid and carbamoyl phosphate, and the remainder may be pooled in cells.

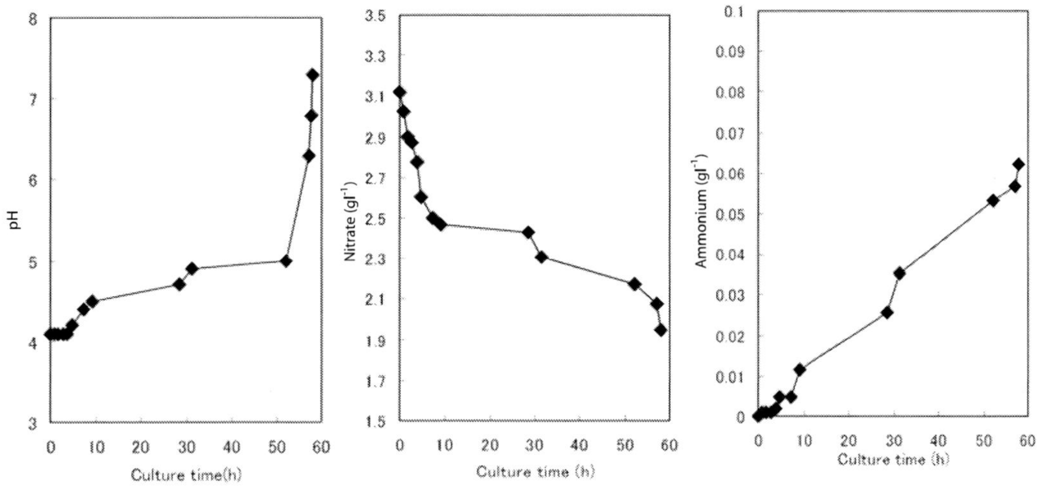

Figure 5. pH change, decrease in nitrate and secretion of ammonium. *R. delemar* NBRC4697 cells cultured in LB medium at pH4.0 with nitric acid.

It might be considered that microorganisms that can degrade nitrate to N2 are promising candidates for bioremediation of acid rain. However, these denitrification reactions are activated only under anaerobic condition, and further acid rain contains many kinds of acids other than nitric acid, such as sulfuric and hydrochloric acids. Therefore, it is difficult to utilize those strains for remediation. Fungi, which can neutralized many kinds of acids under aerobic condition with combination of two neutralization mechanisms, seemed to be desirable species for remediation of acidified soil and water.

4. A Novel Bioremediation Method of Soil Pollution Applied to a Huge Area

In remediation of polluted soil, great attention has been paid to bioremediation because the cost for processing is expected to be lower than those of physical and chemical remediation methods. Degradation of pollutants to carbon dioxide and water by microorganisms is also a desirable merit from the viewpoint of global environment. Operations in bioremediation are classified into two categories, biostimulation and bioaugmentation. In biostimulation microorganisms living in soil are activated (Conner, 1988; Lawes 1991). Supply of oxygen is most effective for the activation. The Bioventing Process, combined with an *in situ* bioremediation process, was often applied to the remediation of leaked fuel in soil. Bioventing process is performed by supplying air (or oxygen) and nutrients to polluted zone (a deep layer of soil) with a high-pressure pump by setting equipments at the polluted place. The running cost and the construction cost for equipment were not low enough, although the costs were lower than those of *ex situ* bioremediation processes (Werner *et al.* 1997, Spain 1994). Several years ago, Oxygen Release Control (ORC) method was developed by REGENESIS company to supply oxygen for a long time (Kunukcu 2007). In this method slurry containing MgO_2 particles was supplied with a high-pressure pump. Activity of microorganisms was maintained for about a half year because oxygen was gradually and continuously released from the slurry by degradation of MgO_2 to $Mg(OH)_2$ and O_2. The ORC method is widely used for *in situ* bioremediation of soil pollutions in the United States. In bioaugumentation, microorganisms, which are screened or genetically produced, are poured into soil and used to decompose harmful chemicals, and thus effective bioremediation methods could be constructed (Kawasumi *et al.* 1984). However, there are few reports of practical application on bioaugumentation.

In acidified soil by acid rain, agricultural chemicals in rice fields and oil lake spilt from oil refineries by accidents, the pollutants spread over a wide range of soil, and in some cases supply of pollutants continues for a long period. In addition, it is required to rapidly and constantly decompose pollutants within a shallow layer in soil to avoid harmful effects on ecosystem and ground water. Because of development of Bioventing and ORC processes, the total costs became much lower than the other remediation methods. However, these bioremediation methods can't be applied to a relatively large area of pollutants in soil. Therefore, a new bioremediation method satisfying these requirements must be developed.

We proposed a novel method, by which microorganisms are immobilized in shallow layer of soil by their self-aggregation (Yasuda *et al.* 2003), because these immobilized microorganisms can decompose target chemicals, or can co-immobilize other microorganisms in the shallow layer for a long period. We screened microorganisms that can form strong aggregates, and *Bacillus subtilis* NBRC3335 was selected from 32 strains. Figure 6 shows the aggregation and retention behavior of *Bacillus subtilis* NBRC3335 in soil and sand. Soil (120 g, average diameter 40 μm) or sand (average diameter 400 μm) were packed in a vessel of 5cm in inside diameter and 5cm in height, and were saturated with distilled water (pH 5.7) at room temperature. The void volume of the packed soil and sand were 63% and 45%, respectively. *Escherichia coli* DH5α cells which showed no aggregation were used as control. *B. subtilis* and *E. coli* cells grown in LB medium were washed with saline and diluted to 0.01 OD (610 nm). The suspension was applied to the surface of soil or sand packed in the vessel, and effluent from the bottom of the vessel was collected. The washing procedure was repeated 10 times at 10-minute intervals. The number of cells in the effluent was measured by counting the number of colonies grown on LB agar plates by incubation for 12 h at 37 °C. As shown in Figure 6, a large amount of *E. coli* cells were passed through the column, especially in the case of the column packed with sand, which had the larger particle diameter and lower water content than those of soil. On the contrary, *B.subtilis* cells were well retrained even in the column packed with sand. This result suggests that *B. subtilis* cells can be used for self-immobilization in soil and sand.

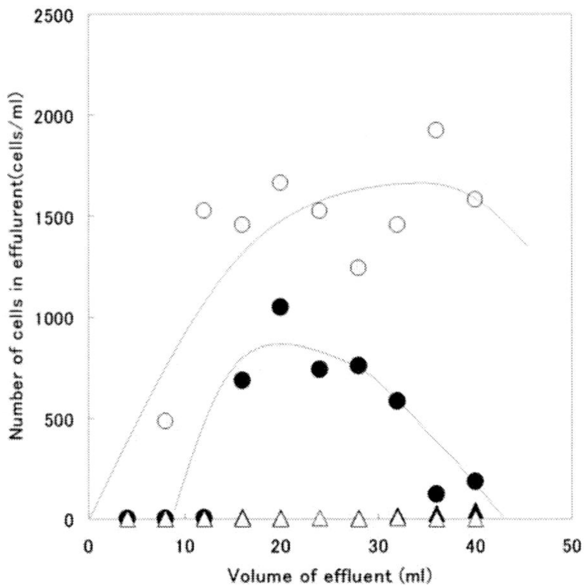

Figure 6. Elution of *B. subtilis* and *E. coli* cells from soil and sand. ○, *E. coli* (sand); □, *B. subtilis* (sand); •, *E. coli* (soil); ▲, *B. subtilis* (soil).

Thus, we proposed a novel bioremediation method named "Bioremediation with Self-immobilized System (BSIS)" to apply the remediation of soil pollution in a large area. The scheme of BSIS is shown in Figure 7. In this method, *Bacillus subtilis* cells are sprinkled on the soil surface at first, and a medium is supplied. *B. subtilis* grows with secretion of viscous polymer, and is self-immobilized by forming cell aggregates in shallow layer (0-5 cm). Viscous polymer filaments are constructed in the shallow layer of soil by this step. Then, cells of a strain, which shows high remediation activity of pollutants, are sprinkled and a medium is again supplied. The cells for remediation are trapped by viscous polymer filaments and can be easily co-immobilized by aggregation with *B. subtilis* in the layer. In case the size of the cells are larger than bacteria, the co-immobilization process can be performed in one step by sprinkling and growing the mixture of *B. subtilis* and the strain for remediation.

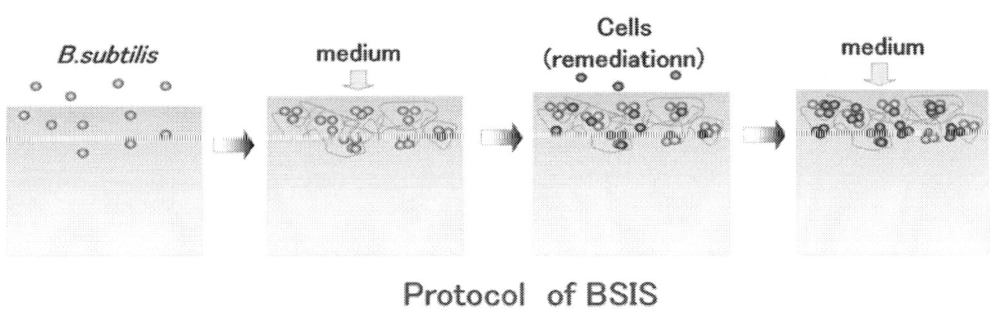

Figure 7. Scheme of BSIS (Bioremediation with self-immobilized system).

BSIS has many advantages in comparison with the other bioremediation method. Most bioremediation methods treat pollutants at a deep layer in soil and thus need high-pressure pumps to supply oxygen and nutrients for activation of microorganisms living in this zone. Cost for the equipment and operation become high, and it is difficult to apply these methods to pollution spread over a large area. However, BSIS needs not to supply oxygen and nutrients with high-pressure pumps, because BSIS treat pollutants at a shallow layer in soil, where oxygen and nutrients are easily supplied from soil surface without use of any expensive equipment. In addition, BSIS can perform self-immobilization of microorganisms by simply sprinkling cells. Bioremediation *in situ* (at polluted site) is possible because of this simple operation, and BSIS can be applied even in case many plants and trees are growing. These advantages suggest that BSIS is applied to pollution spread over a large area.

5. Bioremediation of Acidified Soil with Bioremediation with Self-immobilized System (BSIS)

Maintenance of the pH value at around neutral in soil is important for preservation of microbial ecosystems and plants. It is more desirable if acid rain is neutralized within a shallow layer (so-called A0 and A1 zones) just after falling on soil, because harmful effects, such as leaching of K^+, Mg^{2+} and Ca^{2+}, decrease in root nodule bacteria and contamination of ground water, can be minimized. Thus, performances of BSIS for bioremediation of acidified soil by using fungus NPH6 were studied to achieve an effective remediation (Yasuda *et al.* 2004). Figure 8 shows the experimental apparatus used for neutralization of acids in model soil by BSIS. An acrylic column, 5 cm in inside diameter and 57 cm in height, was packed with sterilized silica sand (600-800 μm of particle sizes) used as a model soil. The height of the packed bed was 40 cm and the void fraction was 0.41. The column was equipped with eight sample ports and a pH electrode was inserted into the center of the column through the second sampling port for measurement of the pH value of flowing liquid at 7.5 cm from the top surface of packed silica. The temperature of the column was kept at around 30 °C.

Self-immobilization at shallow layer with *pullulans* NPH6 and *B. subtilis* was carried out as follows. *A. pullulans* NPH6 and *B. subtilis* NBRC3335 cells were pre-cultured in the LB medium (pH 7.2) for 18 h. No mutual interaction was observed, and the growth rates of *A. pullulans* NPH6 and *B. subtilis* in the mixed culture were almost the same values in a pure culture. These cells were harvested by centrifugation and suspended in 20 ml of fresh LB medium (pH 7.2). Each cell concentration of *A. pullulans* NPH6 and *B. subtilis* NBRC3335 was adjusted at 1.0 OD. After filling the packed column with sterilized water, the cell suspension was applied to the top of the column at a flow rate of 16.6 cm^3/h. Then the LB medium (pH 7.2) was supplied at a rate of 5.6 m/h for 18 h to grow and co-immobilize these cells (mixed culture). For comparison, immobilization was also carried out by application of only *A. pullulans* NPH6 (mono culture).

After immobilization, 50 mM acetate buffer (pH 5.0) was supplied for 18 h at a rate 16.6 ml/h, which corresponds to rain fall of 8 mm/h. Figure 9 (a) shows the distribution of the cell number of *A. pullulans* NPH6, which is expressed as the total cell number in each section of the column, after supply of the medium for 18 h. By supply of LB medium, the cell number in

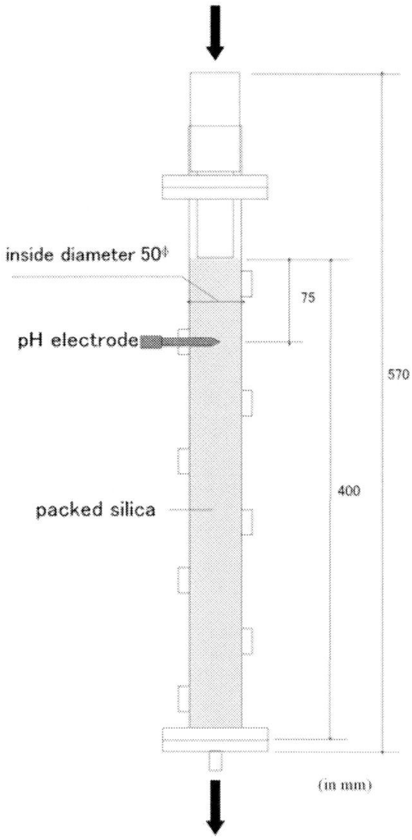

Figure 8. Schematic diagram of the column used for immobilization of cells in BSIS.

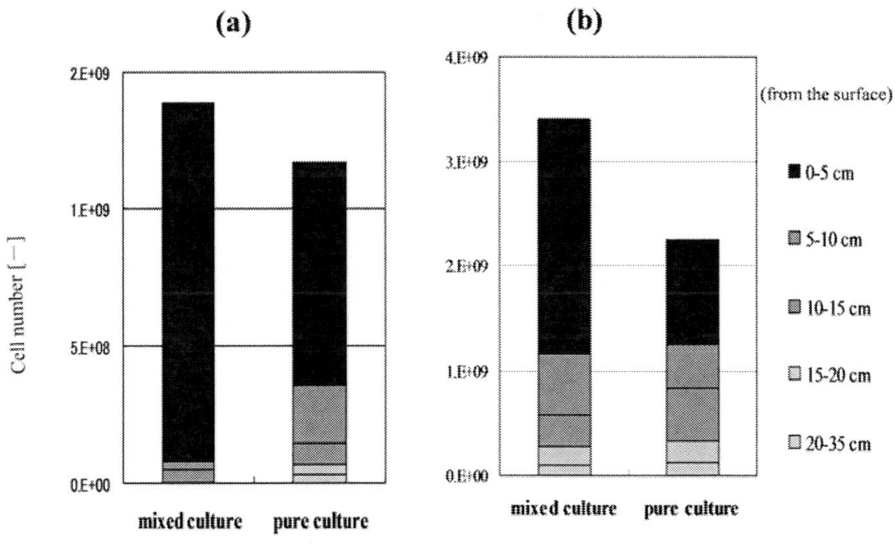

Figure 9. Distribution of the cell number of *A. pullulans* NPH6 along the column axis: (a) before supply of acetate buffer (pH 5.0), (b) after supply of acetate buffer (pH 5.0) for 18 h.

the column increased about 17 times that initially supplied in the mixed culture. This increase in the cell number is almost the same with that in the batch culture. Further, in the mixed culture over 80% of total cells existed in the layer of 0-5 cm from the surface of the packed silica. The DO value measured by a DO electrode inserted to a depth of 7.5 cm was about 50% of air saturation after the supply of LB medium for 18 h. Figure 9 (b) shows the distribution after supplying the 50 mM acetate buffer for 18 h. The cell number became twice, and in the mixed culture 60% cells still existed in the layer. Thus, *A. pullulans* NPH6 was stably immobilized with *B. subtilis* in a shallow layer, where sufficient oxygen for aerobic growth was supplied.

Figure 10 shows the pH value at 7.5 cm and 40 cm from the surface, together with the cell numbers in the effluent solution from the column, when the acetate buffer was supplied. Without inoculation of cells (control) the pH values approached to about 5, which is the initial value of the applied acetate buffer. At 7.5 cm from the surface, the pH value of the flowing liquid in the mixed culture was kept around 7 for 18 h, but it decrease below 6 in the pure culture after 12 h. In the pure culture the number of the cells in effluent was larger than that in the mixed culture and increased rapidly after 12 h. These results show that *B. subtilis* could effectively immobilize *A. pullulans* NPH6 and could maintain *A. pullulans* NPH6 in the shallow layer of the packed silica. Figure 11 shows neutralization characteristics of the sulfuric acid solution by BSIS. The pH value in sulfuric acid solution was kept above 6.5 for 96 hours, while it decreased below 5 in pure culture. The total amount of the solution corresponded to the average value of rainfall of one year in Japan. This means that BSIS is applicable to neutralization of acidified soil caused by acid rain for a relatively long time.

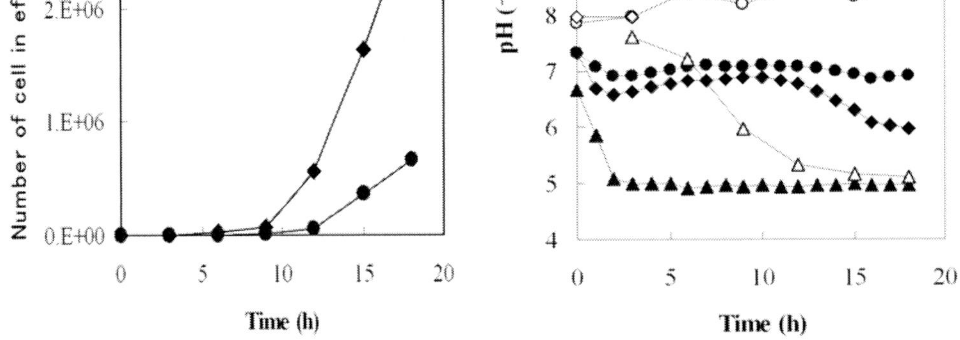

Figure 10. Cell number of *A. pullulans* NPH6 in effluent and pH value: (a) cell number in effluent (at 40 cm from the top of packed silica), (b) pH value at 7.5 cm and 40 cm from the surface.

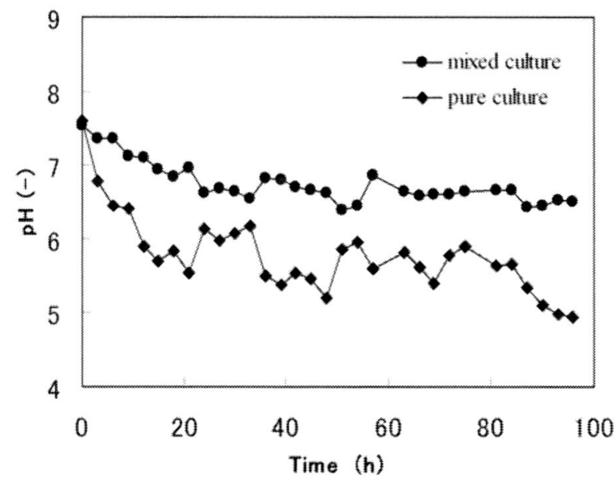

Figure 11. Neutralization of sulfuric acid with *A. pullulans* NPH6 in mixed culture and a pure culture (pH value at 7.5 cm).

6. Neutralization of Acid in Lake Water With Fungi

We also tried to develop a bioremediation method for acidified water in lakes and ponds with fungi *R. delemar* NBRC4697 (Nabeshima *et al.*). Pond water used as model water, which contained about 10^3 cells/ml of indigenous microbial cells, was obtained from a pond located at Kozubashi (Nishi-ku, Kobe, Hyogo, Japan). To examine the neutralization ability of indigenous microorganisms and *R. delemar*, 10 ml of model pond water (pH 7.8) were adjusted to pH 4.0 or 5.0 with nitric acid and incubated without sterilization at 30°C. Figure 12 shows the pH change, when the pond water with and without *R. delemar* NBRC4697 was incubated at 30°C. The indigenous microorganisms could increase the pH value of the pond water at pH 5.0 but could not neutralize the pond water at pH 4.0 because of their low acid neutralization abilities. When *R. delemar* NBRC4697 cells were added to the pond water, the pH value increased from 4.0 to 7.0 within 2 days. To clarify the effect of *R. delemar* cells on the growth of other bacteria, 0.2 ml of pre-cultured *E. coli* cells, a model bacterial strain, were incubated in 10 ml of sterilized pond water (no indigenous cells) at pH 4.0 with and without 0.25 mg of *R. delemer* NBRC4697 wet cells. As shown in Figure 13, *E. coli* cells showing a low ability to neutralize acid did not increase pH, and did not grow without *R. delemar* cells (open squares) at pH 4.0. When *R. delemar* cells (close squares) were added, the pH value of the pond water increased gradually, and *E. coli* cells grew as a co-culture with fungi cells. The result suggest that neutralization of acidified pond and lake water by adding *R. delemar* is suitable, because suitable conditions for the growth of indigenous microbes can be recovered.

Although *R. delemar* NBRC4697 cells are useful for neutralizing acidified pond water, their ability was not high enough to neutralize a large amount of water in a short period of time. Thus, an immobilization method of cells using a seamless cellulose tube (36 mm in width and 50 mm in length) was studied for the repeated use of *R. delemar* as a neutralizer.

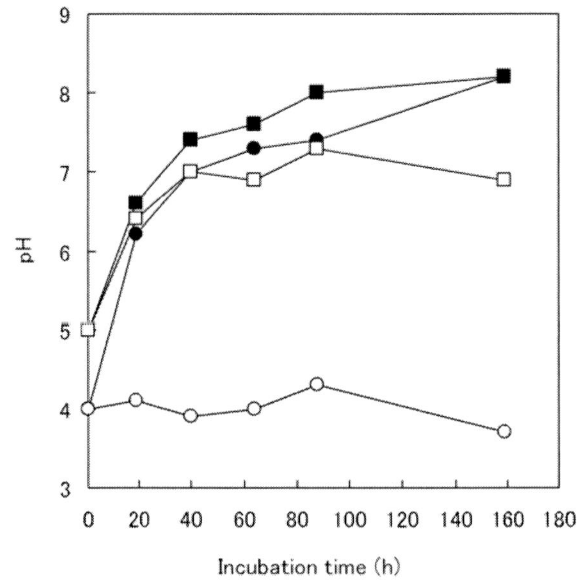

Figure 12. Acid neutralization with indigenous microorganisms and *R. delemar* NBRC4697 in pond water adjusted at pH 4.0 and 5.0 with nitric acid. Circles, pH 4.0; Squares, pH 5.0; solid keys, pond water with *R. delemar*; open keys, pond water without *R. delemar*.

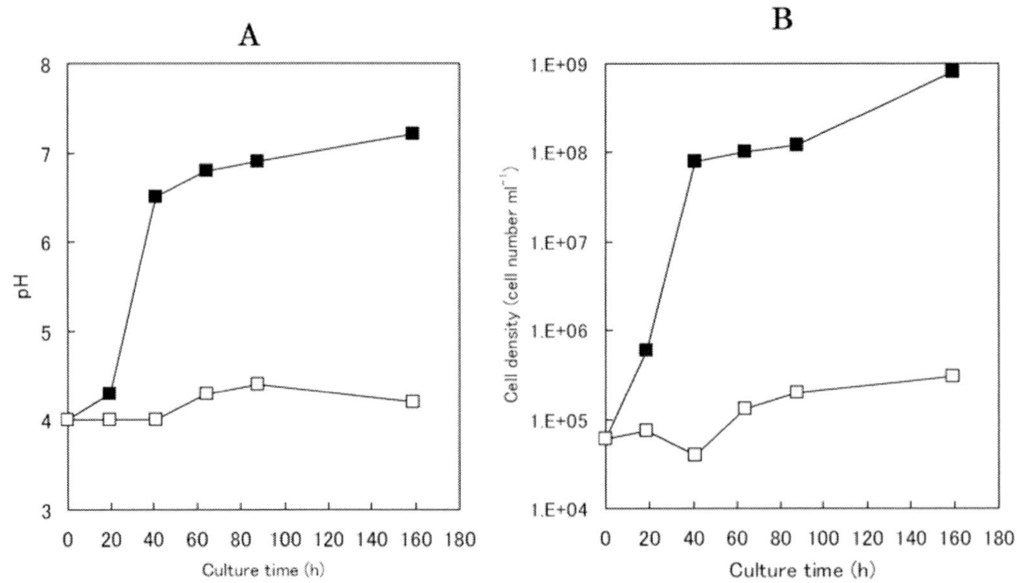

Figure 13. Effect of increase in pH by *R. delemar* NBRC4697 cells on the growth of *E. coli* cells. (A) pH of pond water; (B) cell densities of *E. coli*. Solid squares, co-culture of *E. coli* and *R. delemar*, open squares, mono-culture of *E. coli*.

Low molecular materials, such as NO_3^- and NH_4^+, can permeate the cellulose tube, but cells and spores cannot. The *R. delemar* cells cultured in 50 ml of LB medium was suspended in 5 ml of pond water. The cell suspension was added to the cellulose tube and incubated in 200 ml of pond water adjusted to pH 4.0 with nitric acid at 30°C without shaking. The tube contained air, and a part of the tube was outside the pond water. During a 4-day period, the pond water was replaced with fresh pond water (pH 4.0) four times using the draw-fill method. The concentrations of NO_3^- and pH value were measured during cultivation. Figure 14 shows the change in pH and the concentration of nitrates in pond water. Although the volume ratio of the cell suspension to pond water was only one-fortieth, the pH value increased from 4.0 to 7.8 within 1 hour, and nitrates decreased below 0.02 g/l. Moreover, during the four times of the medium change, no decline in the neutralization abilities was observed. These results suggest that the immobilized *R. delemar* cells can continuously neutralize pond water and this immobilization method is useful in neutralizing pond water for a long period.

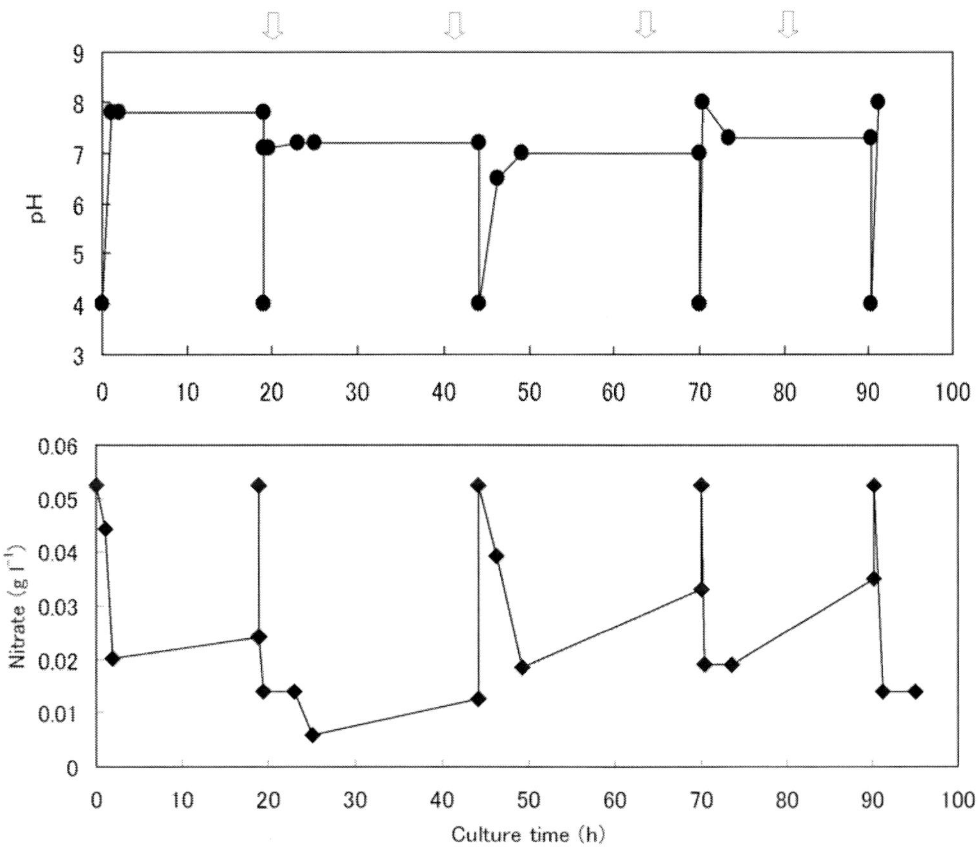

Figure 14. Change in pH and nitrate in draw-fill method with immobilized *R. delemar* NBRC4697. Arrows, change of pond water; volume ratio of cell suspension to pond water was 1:40.

7. Advantages and Problems in Bioremediation with Fungi in Practical Application

The bioremediation method BSIS has some advantages in the remediation of acid rain. We developed the BSIS method as an alternative of spreading of lime (so-called liming) and showed neutralization of acids in model soil. The immobilized fungi with BSIS can neutralize more effectively and for longer time than the liming because of rapid neutralization at shallow layer and stable co-immobilization of cells. To prepare a layer of immobilized cells in BSIS, cells and medium were spread on soil. This procedure seems similar to the liming procedures. Therefore, BSIS may also be used simultaneously with the liming for the remediation of acidified soil.

Moreover, since cells are well co-immobilized with *B. subtilis*, many kinds of cells, which have abilities to degrade pollutants, can be easily immobilized, and thus this method will be widely used for bioremediation. For example, triazine herbicides, which are used for farms and are suspected to be possible carcinogens and endocrine disrupters, must be degraded rapidly after usage, because excess use of triazine herbicides frequently contaminates soil and groundwater. We immobilized *Pseudomonas putida* NRRL B12227 at shallow layer by BSIS to degrade cyanuric acid, an intermediate of degraded triazine herbicides. More than 70% of the *Pseudomonas* sp. NRRL B-12227 cells were co-immobilized with the *B. subtilis* in a 7.5 cm layer of the packed soil by self-aggregation. More than 60% of the 1 mM cyanuric acid supplied to the packed soil was degraded in this layer during a 72 h period (Shiomi *et al* 2006).

Several problems to be solved are still remaining for practical application in bioremediation by fungi. First point is the activity of fungi strains for neutralization of acids. Fungi strains show high activities for neutralization compared with other species, but in practical application the activity must be enhanced much more by screening and/or mutation. Adaptation of the strains to environmental conditions, especially temperature, is also an important factor to be considered. Most areas damaged by acid rain are located in northern areas of Europe and Canada, and under low temperature the microorganisms may not actively grow. Since many mutants resistant to low temperature and freezing in *Sccharomyces cerevisiae* have been obtained to produce frozen dough, screening of mutants from fungi will be practicable to obtain strains having high remediation activity under low temperature. Genetic methods, if application of recombinant strains to field tests is allowed, seem to be more promising to obtain strains with high activity.

Second point to be considered is synergy effect of indigenous ecosystems on remediation by fungi. For example, some bacteria, such as *Ntrosomonas europaea* and *Nitrobactor winogradskyi*, change ammonia to nitrates by nitrification (Bothe *et al.* 2000, Ilies and Mavinic 2001) and as a result, they decrease the pH value by use of ammonia secreted from fungi. As far as our results, however, soil microorganisms, which were collected from many soil samples, assimilated ammonium ion as a nitrogen sources but did not produce nitrates. This result shows that strains converting ammonium ion to nitrates were few among soil microorganisms close to soil surface. However, soil and water contain many other indigenous microorganisms, and their ecosystems are very complicated. Therefore, the relationship between fungi and indigenous microorganisms has to be carefully studied.

It is reported that algae, seaweeds and mosses also decrease nitrates and ammonium ion by absorption. Complicated mechanisms have been studied by monitoring acidified lake water. The chemicals produced by photochemical, chemical and biological reactions increased alkalinity in water (Kopacek *et al.* 2003, Kerekes and Freedman 1989, Kleeberg *et al.* 2006). Photochemical reactions by phytoplanktons, green algae, diatom and macrophyte were important factors decreasing dissolved organic anions and thus increase alkalinity, and the clarity of water affect these photochemical reactions (Kerekes and Freedman 1989). The combination of those strains with the fungi strains showing neutralization ability may expect synergy effects to maintain pH value at neutral. Moreover, acid sensitive species are replaced by acid tolerant species in acidified lakes (Shindler 1988). Indigenous strains in such acidified lakes may be resistant to acid, and addition of fungus cells for bioremediation might unexpectedly disturb these ecosystems. The effect on these indigenous ecosystems must be clarified before application.

The third point is design of equipment and procedures for practical application. In a bioremediation of soil, we proposed BSIS method and this method needs not special devices for oxygen and nutrient supply. However, in order to apply this BSIS over a large area, spraying methods of cells and nutrients have to be developed. For example, if a higher cell density in shallow layer in soil is required to increase remediation efficiency, solid nutrients, which are gradually melted and released in soil for a long time, have to be supplied from soil surface. The process design to the remediation of lake and pond water is also necessary, although fungi cells immobilized in a cellulose tube with the volume ratio 1: 40 could be used repeatedly for several days, as shown in Figure 14. Acidic water flows into and out lakes, and for example, lakes in Norway are rapidly acidified by the phenomenon called "acid shock" in spring season, which is caused by huge amount acid contained in snow. These suggest that neutralization of water at river mouths to lakes seems effective. In the case of Tazawa Lake, acidified water was neutralized with lime before flowing into the lake. The pH value of Tazawa Lake was increased by this treatment.

The legal regulation is another issue to be considered. EU and FDA regulate scheme of bioaugumantation, and bioaugumantation must be practiced according to the regulation. In the regulation, polluted place has to be registered before approval. Therefore, bioaugumantation over a huge area or unfixed area might not be approved.

As pointed out in this section, many problems for practical application of bioremediation with fungi are still remaining. However, abilities of fungi in bioremediation of acidified soil and water are useful for keeping suitable environment for living systems.

References

[1] Berg, N. H., Gallegos, A., Dell, T., Frazier, J., Procter, T., Sickman, J., Grant, S., Blett, T., & Arbaug, M. (2005). A screening procedure for identifying acid-sensitive lakes from catchment characteristics. *Environ. Monit. Assess.*: **105**, 285-307.

[2] Bothe, H., Jost, G., Scloter, M., Ward, B. B., & Witzel K-P. (2000). Molecular analysis of ammonia oxidation and denitrification in natural environments. *FEMS Microbiol. Rev.*: **24**, 673-690.

[3] Conner, J. R. (1988). Case study of soil vending. *Pollut. Eng.*: **7**, 74-78.

[4] Cotter, P. D., & Hill, C. (2003). Surviving the acid test: responses of gram-positive bacteria to low pH. *Microbiol. Mol. Biol. Rev.:* **67**, 429-453.

[5] Darland, G., Brock, T. D., Samsonoff, W., & Conti, S. F. (1970). A thermophilic acidophilic mycoplasma isolated from a coal refuse pile. *Science*: **170**, 1416-1418.

[6] Elferrink, M. G. L., de Wit, J. G., Demel, R., Driessen, A. J. M., & Konings, W. K. (1992). Functional reconstitution of membrane proteins in monolayer liposomes from bipolar lipids of *Sulfolobus acidocaldarius*. J. Biol. Chem.: **267**, 1375-1381.

[7] Evans, J. S. (1982). Biological effects of acidity in precipitation on vegetation. A Review. *Environ. Exper. Bot.*: **22**, 155-169.

[8] Forsius, M., Vuorenma, J., Mannio, J., & Syri, S. (2003). Recovery from acidification of Finnish lakes: regional patterns and relation to emission reduction policy. *Sci. Total Environ.*: **310**, 121-132

[9] Hirata, T., Kaneko, T., Ono, T., Nakazato, T., Furukawa, N., Hasegawa, S., Wakabayashi, S., Shigekawa, M., Cang, M. H., Romero, M. F., & Hirose, S. (2003). Mechanism of acid adaptation of a fish living in a pH 3.5 lake. *Am. J. Phys. Reg. Int. and Com. Phy.*: **284**, 1199-1212.

[10] Ilies, P., & Mavinic, D. S. (2001). A pre- and post-denitrification system treating a very high ammonia landfill leachate: effects of pH change on process performance. *Environ. Technol.*: **22**, 289-300.

[11] Jeffries, D. S., Brydges, T. G., Dillon, P. J., & Keller, W. (2003). Monitoring the results of Candida/U.S.A. acid rain control programs; some lake responses. *Environ. Monit. Assess.*: **88**, 3-19.

[12] Kawasumi, T., Igarashi, Y., Komada, Y., & Minoda, K.Y. (1984). *Hydrogenbactor thermophilus gen nov.*, an extremely thermophilic aerobic hydrogen-oxidizing bacterium. *J. Syst. Bacteriol.*: **34**, 5-10

[13] Keller, W., Heneberry, J. H., & Dixit, S. S. (2003). Decreased acid deposition and the chemical recovery of Killaney, Ontario, lakes. *Ambio*: **32**, 183-189.

[14] Kerekes, J., & Freedman, B.(1989). Characteristics of tree acidic lakes in Kejimkujik National Park, Nova, Scotia, Canada. Arch. Environ. Contam. *Toxicol.*: **18**, 183-200.

[15] Kleeberg, A., Schubert, H., Koschorreck, M., & Nixdorf, B. (2006). Abundance and primary production of filamentous green algae *Zygogonium ericetorum* in an extremely acid (pH 2.9) mining lake and its impact on alkalinity generation. *Freshwater Biol.*: **51**, 925-937.

[16] Kopacek, J., Hejzlar, J, Kana, J., & Porcal, P. (2003). Photochemical, chemical, and biological transformations of dissolved organic carbon and its effect on alkalinity production in acidified lakes. Limnol. *Oceanography*: **48**, 106-117.

[17] Kunukcu, Y. K. (2007). In situ bioremediation of groundwater contaminated with petroleum constituents using oxygen release compounds (ORCs). Environ. Sci. Health A TOx. Hazard Subst. *Environ. Eng.*: **42**, 839-45.

[18] Larssen, T., Brereton, C., & Gunn, J. M. (2003). Dynamic modeling of recovery from acidification of lakes in Killarney Park, Ontario, Canada. *Amobio*: **32**, 244-248.

[19] Lawes, B. C. (1991). "Soil-induced decomposition of hydrogen peroxide" In-situ bioremediation: application and investigation for hydrocarbon and contaminated site remediation, R.E.Heichee and R.F. Offenbuttel, eds. Pp143-156, Butterworth-heinemann Stoneham, Boston, USA (1991)

[20] Materechera, S. A., & Mkhabela, T. S. (2002). The effectiveness of lime, chicken manure and leaf litter ash in ameliorating acidity in a soil previously under black wattle (Acasia mearnsii) plantation. *Bioresour. Technol.*: **85**, 9-16.

[21] Matzner, E., Khanna, P. K., Meiwes, K. J., & Ulrich, B. (1983). Effects of fertilization on the fluxes of chemical elements through different forest ecosystems. *Plant and Soil*: **74**, 343-358.

[22] Nabeshima, R., Kusumoto, N., Katoh, S., & Shiomi, N., Neutralization of acidified water by use of Rhizopus delemar immobilized with a cellulose membrane. Submitted.

[23] Nguyen, V.A., Senoo, H., Misima, K., & Hisamatsu, M. (2001). Multiple tolerance of *Rhodotorula glutinis* R-1 to acid, aluminum ion, and its usual ability of neutralizing acidic medium. *J. Biosci. Bioeng.*: **92**, 366-371.

[24] Schindler, D. J. (1988). Effects of acid rain on freshwater ecosystems. *Science*: **239**, 149-157.

[25] Searcy, D. G. (1976). *Thermoplasma acidophilum*: intracellular pH and potassium concentration. *Biochim. Biopys. Acta:* **451**, 278-286.

[26] Seoane, S., & Leirós (2001). Acidification-neutralization process in a lignite mine spoil amended with fly ash or limestone. *J. Envion. Qual.* **30**, 1429-1431.

[27] Shiomi, N., Yasuda, T., Inoue, Y., Kusumoto, N., Iwasaki, S., Katsuda, T., & Katoh, S. (2004). Characteristics of neutralization of acids by newly isolated Fungi cells. *J. Biosci. Bioeng.*: **97**, 54-58.

[28] Degradation of cyanuric acid in soil by *Pseudomonas* sp. NRRL B-12227 using bioremediation with self-immobilized system. *J. Biosci. Bioeng.* **102**, 206-209.

[29] Sienkiewicz, E., Gasiorowski, M., Hercman, H. (2006). Is acid rain impacting the Sudetic lakes? *Sci. Total Environ.*: **369**, 139-149.

[30] Snucins, E., Gunn, J., Keller, B., Dixit, S., Hindar, A., & Henriksen, H (2001). Effects of regional reductions in sulfur deposition on chemical and biological recovery of lakes with Killary Park, Ontario, Canada. *Environ. Monit. Assess.*: **67**, 179-194.

[31] Spain, J. C. (1994). Environmental biotechnology research: an overview. *Aviat. Space. Environ. Med.*: **65**, A131-137.

[32] Stevens, G., & Dunn, D. (2004). Fly ash as a liming material for cotton. *J. Environ. Qual.*: **33**, 343-348.

[33] Sverdrup, H., Martison, L., Alveteg, M., Moldan, F., Kronnas, V., & Muthe, J. (2005). Modeling recovery of Swedish ecosystems from acidification. *Ambio*: **34**, 25-31.

[34] Swenson, W. A., McCormick, J.K., Simonson, T. D., Jensen, K. M., & Eaton, J. G. (1989). Experimental acidification of Little Rock Lake (Wisconsin): fish research approach of early responses. *Ach. Environ. Contam. Tox.:* **18**, 167-174.

[35] Takaya, N. (2002). Review: Dissimilatory nitrate reduction metabolisms and their control in fungi. *J. Biosci. Bioeng.*: **94**, 506-510.

[36] Tipping, B., Bass, J. A., Hardie, D., Haworth, E. Y., Hurley, M. A., & Wills, G. (2002). *Environ. Poll.*: **116**, 137-146.

[37] Tammi, J., Appelberg, M., Beier, U., Hesthagen, T., Lappalainen, A., & Rask, M. (2003). Fish status survey of Nordic lakes: effects of acidification, eutrophication and stocking activity on present fish species composition. *Ambio*: **32**, 98-105.

[38] Vrba, J., Kopacek, J., Fott, J., Kohout, L., Nedbalova, L., Prazakova, M., Soldan, T., & Schaumburg, J. (2003). Long-term studies (1871-2000) on acidification and recovery of lakes in the Bohemian Forest (central Europe). *Sci. Total Environ.*: **310**, 73-85.

[39] Warby, R. A., Johnson, C. E., & Driscoll, C. T. (2005). Chemical recovery of surface waters across the northeastern United States from reduced input of acidic deposition: 1984-2001. *Environ. Sci. Technol.*: **39**, 6548-6554.

[40] Werner, F. T., Walters, J. E., & Keefer, G. B. (1997). Bioventing pilot test results at the low point drain area, Offutt AFB, Nebraska. *Ann. N. Y. Acad. Sci.*: **829**, 313-325

[41] Yasuda, T., Katoh S., Inoue, Y., & Shiomi, N. (2003). A bioremediation method based on self-immobilization of cells in shallow layer of soil. *J. Chem. Eng. Jpn.:* **36**, 216-219.

[42] Yasuda, T., Shiomi, N., Iwasaki, S., Yamaguchi, Y., Katsuda, T., & Katoh, S. (2004). Neutralization of acids by microorganisms co-immobilized with *Bacillus subtilis* in a shallow layer of model soil. *J. Chem. Eng. Jpn.*: **37**, 1445-1451

[43] Zhou, Z., Takaya, N., Nakamura, A., Yamaguchi, M., Takeo, K., & Shoun, H. (2002). Ammonia fermentation, a novel anoxic metabolism of nitrate by fungi. *J. Biol. Chem.:* **277**, 1892-1896.

In: Acid Rain Research Focus
Editor: Lyman G. Roglesfield, pp. 125-142
ISBN: 978-1-60456-373-3

Chapter 5

ACID RAIN HARMS WORLD HERITAGE BUILDINGS AND HUMAN HEALTH IN VIETNAM

***Kazue Tazaki*[1], *Teruhiko Kido*[2], *Naoko Baba*[1] *and Sachie Suzuki*[3]**

[1]Graduate School of Natural Science and Technology, Kanazawa Univ., Kanazawa, Ishikawa, Japan

[2]Medical School of Health Science, Kanazawa Univ., Kanazawa, Ishikawa, Japan

[3]Ishikawa Prefecture Nursing University, Nakanuma, Kahoku City, Ishikawa, Japan

Abstract

Acid rain's negative impact on world heritage buildings and human health in Vietnam is not very well known. Historical monuments, statues, and buildings are covered by black soot and fly ash, caused by fossil fuels in acid rain in Hue, Cam Phuc, and Cam Chinh in Central Vietnam. The sooty surfaces of study subjects show damage by not only acid rain, but also by fungi, lichens, and bacteria. Rainwater collected in January 2006 and August 2007 in 3 villages and observed by electron microscopy showed abundant carbon spherules and fly ash, associated with clay minerals, iron materials and microorganisms. Most Vietnamese drink well water and/or collected rainwater, especially in the countryside. Sometimes acid rain water shows a pH of 6-8, because of the effects of sea salts, but its electrical conductivity (EC) remains quite high (eg:300-1200 μ S/cm), suggesting high concentrations of SOx and NOx components. Chemical, biological, and environmental studies were performed on world heritage buildings in Dong Ha, whereas the rain and drinking water collected in Hanoi, Cam Chinh (30 homes) and Cam Phuc (32 homes) were studied, in association with concurrent health check interviews. Some heavy metals (Cr, Fe, Zn, Br, and Sr) and colon bacilla were found in the drinking water even after filtrations at Cam Phuc Village. In Cam Phuc, incidence of diseases such as anemia, roundworm, dermatitis, and symptoms of fatigue, loss of appetite, skin problems, stomachache, and dizziness is worse than that in Cam Chinh, indicating no correlation between former Agent Orange dispersion and current health conditions covered by this study. The worst health conditions were caused by drinking water in Cam Phuc village, a control area with no history of Agent Orange contamination. Further more, volatile organic compound concentrations of formaldehyde, acetaldehyde, acetone, propionaldehyde, hexaldehyde, and 2,5-dimethylbenzaldehyd in both Cam Chinh and Cam Phuc villages were quite high. Especially high acetone concentrations (13-59 μ g/m^3) were detected both indoors and outdoors, and showed more than 100 times the acetone levels of the control standard in Japan. The origin of the acetone is unknown.

Keywords: acid rain, fly ash, world heritage building, drinking water, health condition, heavy metals, volatile organic compound, Vietnam

1. Introduction

Throughout the world it is known that the U.S. Military used Agent Orange and other defoliants containing dioxins in southern Vietnam during the Vietnam War. The extent and distribution of herbicides is estimated to be several million liters more than the results published by the US National Academy of Sciences in 1974. In many studies, breast milk and serum analyses indicated persistant human health effects on Vietnamese people (Kido et al., 2007; Tawara et al., 2007; Dung et al., 2007; Sawano et al., 2007; Thom et al., 2007; Nishijo et al., 2007). Historical studies were performed on dioxins in human breast milk and their relation with biochemical indicators in serum and health data of infants, adipose tissues, and serum in Cam Chinh commune in Quang Tri province where herbicides were sprayed, and Cam Phuc commune in Ha Tinh province as a control area. It is very difficult to show individual exposure from the dioxin concentrations, but it is still important to clarify individual concentrations of internal dioxins in order to investigate their health effects.

On the other hand, acid rain and drinking water problems in Vietnam are much less studied than the impacts of Agent Orange, even though the International Year of Planet Earth-Global Launch Event (IYPE) will support research projects of "Groundwater" and "Earth and Health". Climate air conditions and the characteristics of precious natural rainwater need to be recorded now. According Science news, wood, stone, glass, and other materials are damaged by climate-influenced factors. For example, in Europe attacks by wood-destroying fungi may increase because of warmer, wetter weather. Humidity has already cracked a medieval wooden altarpiece (Cho, 2007). Ancient historical treasures and human health are clearly being threatened by human induced climate change and other environmental problems.

In this study, more attention has been paid to the impact of microorganisms in black acid rain harming World Heritage treasures and human health in Vietnam, than on dioxin effects. The chemical and biological tests were performed on the world historical buildings, rainwater, drinking water, and volatile organic compounds in the air, which impact the health conditions in villages of Central Vietnam, such as Cam Phuc, Cam Chinh, and Hue (Figure 1).

Viet Nam map of study area and sampling points.

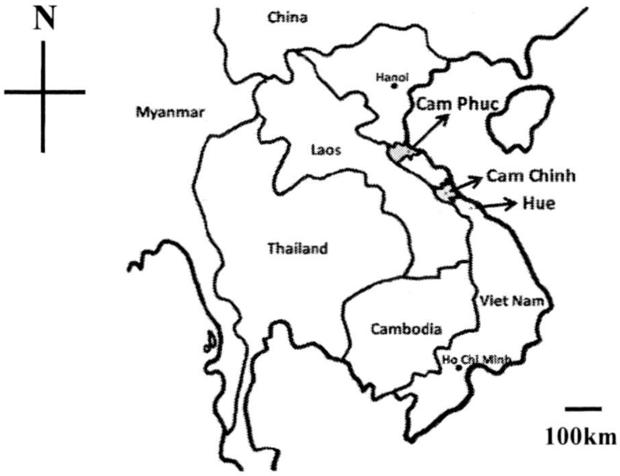

Characteristics of the rain water in Cam Phuc.

Sampling points	pH	Eh (mV)	EC (□S/cm)	DO (mg)	DO (%)	WT ( )
Hotel	**3.4**	**553**	**350**	**9.0**	**96**	**15.3**
Clinic	**6.7**	**218**	**130**	**15.3**	**149**	**12.7**

Measured on January 23, 2006

Figure 1. Abundant black materials were found on the surface of buildings, walls, and beach sand in Hue (A,B) and Cam Phuc (C,D). Arrows indicated black soot areas.

2. Materials and Methods

2.1. Materials

Rainwater and drinking water analyses and concurrent human health checks were performed in two villages in central Vietnam: Cam Chinh commune in Quang Tri province where herbicides were sprayed (abbreviate Quang Tri to QT) and the control area, Cam Phuc commune in Ha Tinh province (abbreviate Ha Tinh to HT), where herbicides were not sprayed during the War (Figure 1). These two communes were separated by the demilitarized zone of 17 degrees north latitude, which was a military boundary in the Vietnam War. A large number of herbicides were sprayed on QT in the south, but none were sprayed on HT in the north.

For comparison, drinking water and rainwater were randomly collected from 50 families in each village on January 2006 and August 2007. Health checks were carried out through interviews of each family using interpreters and Vietnamese questionnaires. Families were asked about health conditions in areas such as anemia, roundworm, dermatitis,symptoms of fatigue, loss of appetite, skin problems, stomachache, and dizziness. The rainwater was filtered and the remains were observed by both optical and electron microscopy. Analysis was done in the field through pH, Eh, EC, DO, water temperature, chemical pack tests, and colon bacillus cultivation.

Dust and rainwater samples were collected from blocks of a collapsed wall at the World heritage church building in Hue, Central Vietnam (Figure 1A and B). The samples contained lichens and fungi and the remaining walls are also covered by lichen and fungi associated with black soot. The exterior wall of a house at Cam Phuc village was stained by black soot (see arrow, Figure 1C). The black soot was washed off of buildings by rainwater to the sandy beach areas leaving black trails (Figure 1D, arrows). Stains on walls and buildings could be seen during the dry season of January 2006, but the black soot went away during the rainy season of August 2007.

2.2. Methods

Chemical analyses were carried out by a JEOL JSX-3600 ED-XRF equipped with an X-ray probe scanning system which was operated using Rh Kα radiation under 50 kV acceleration voltage. An ICP mass spectrometer (VARIAN VISTA-MPX) and an Atomic absorption spectroscope (Thermo-fisher S-4) were used to analyze trace elements in the water chemistry.

The dust samples were observed with a JEOL JEM 2000EX transmission electron microscope (TEM) operating at 120 kV in bright-field mode. A drop in suspension was mounted on a micro grid. After air-drying, the grids were observed without staining for mineralogical investigation. The mineralogical properties of the samples were analyzed, using a Rigaku RINT 1200 X-ray powder diffract-meter (XRD) with Cu Kα radiation, generated at 40 kV and 30 mA. Concurrent health checks for anemia, roundworm, dermatitis, and symptoms of fatigue, loss of appetite, skin problems, stomachache, and dizziness were performed by interviews through a Vietnamese interpreter.

3. Results

3.1. World Heritage Buildings

On January 23, 2006, the rainwater in Cam Phuc indicated strong acid rain with a pH of 3.4 at the hotel whereas the samples taken at the clinic showed a pH of 6.7. The results showed a strong correlation between EC values and pH (Figure 1), suggesting that the pH depends on the amount of black soot. The surfaces of the buildings have been impacted by the rainwater quality.

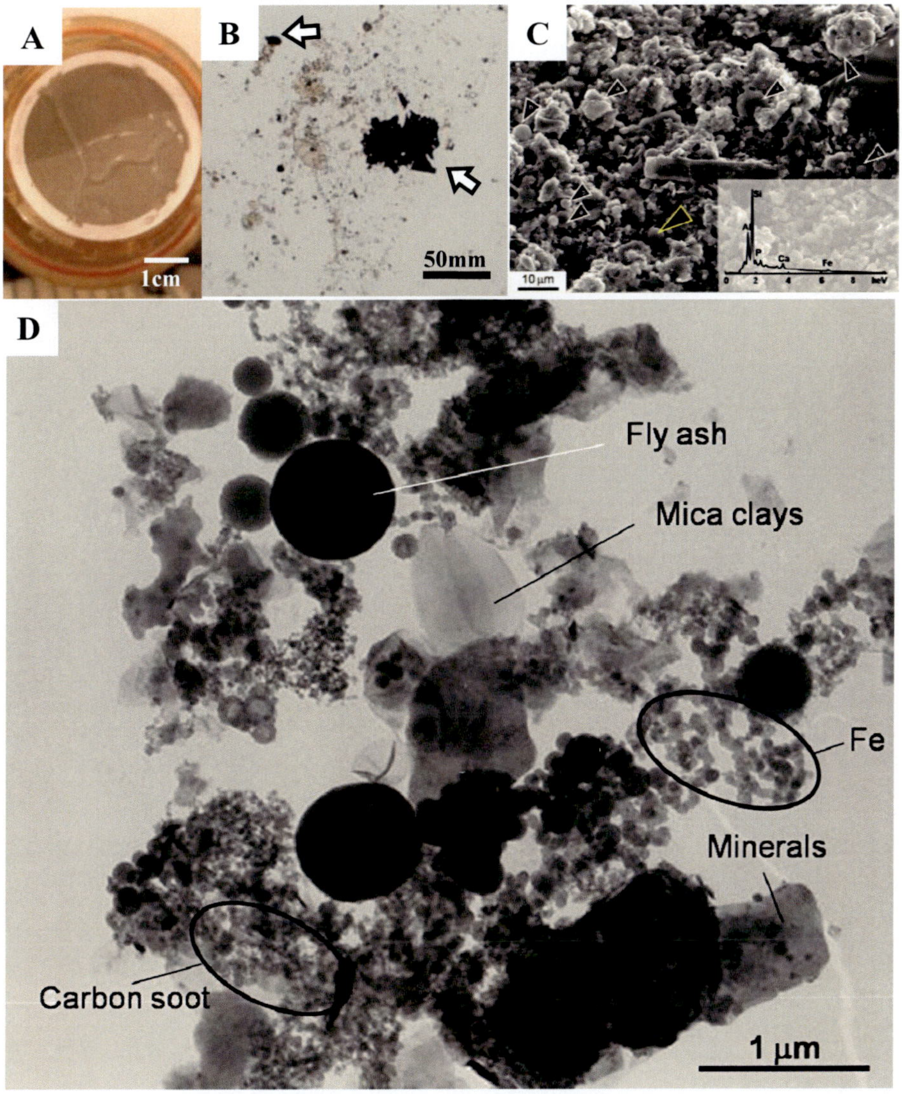

Figure 2. Filtered rainwater in Cam Phuc, showing reddish brown materials (A) contain high-density opaque materials (B, arrows). SEM observation and the EDX analysis of suspension indicated the analytical points showing Si and Al contents (C, arrows, inset). TEM micrograph showed large amount of spherical particles of fly ash, mica clays, Fe-materials, mineral grains, and carbon soot, originated from burning coal (D).

The rainwater was directly collected from the air and compared with materials on the black wall. Optical micrograph observations of filtered rainwater in Cam Phuc (A in Figure 2), showed reddish brown materials and high-density opaque materials (arrows in Figure 2B), whereas the Scanning Electron Micrograph showed aggregates of clay minerals composed of Al and Si (1:2) which can be identified as kaolin minerals (arrows and inset in Figure 2C). The transmission electron micrograph showed more precise micro morphology of the contents such as fly ash, mica clays, Fe granules, mineral grains, and carbon soot, respectively (arrows in Figure 2D).

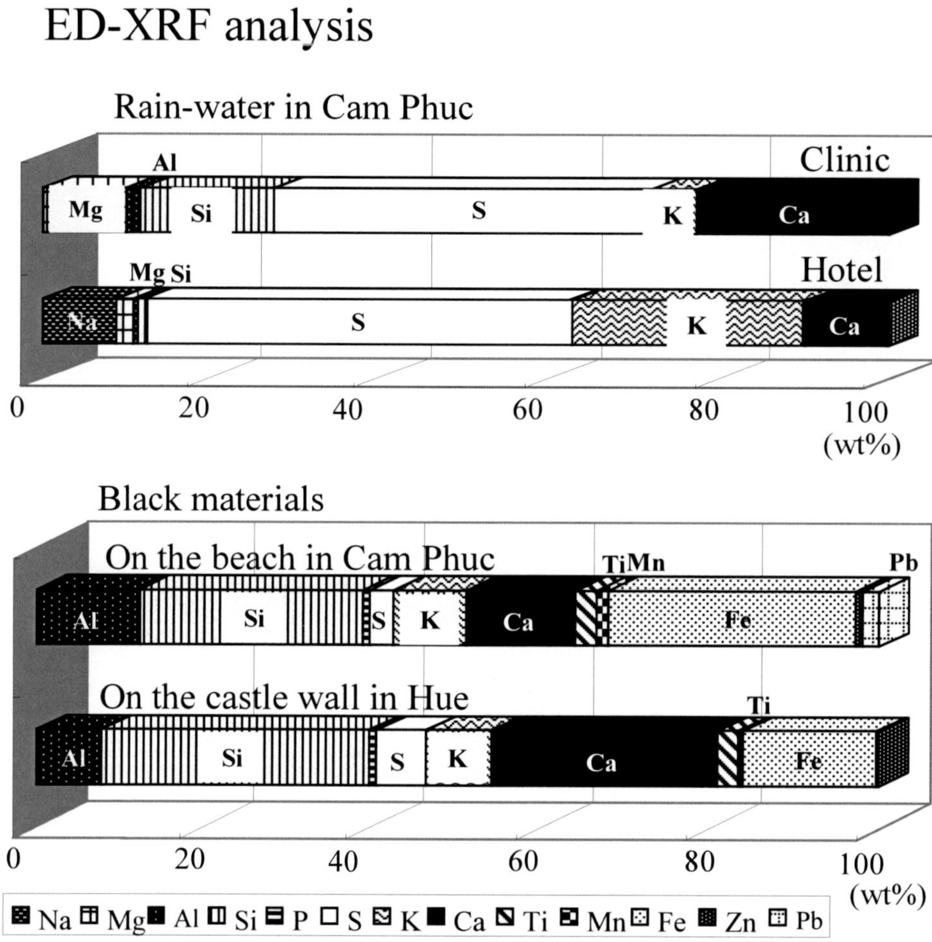

NCS analysis

	N	C	S
Black materials on the beach in Cam Phuc	1.49	46.23	0.21
Black materials on the castle wall in Hue	2.92	31.27	0.46

(wt%)

Figure 3. The ED-XRF analysis indicated high Sulfur content (50 wt%) in rainwater. Both black materials contain high carbon soot, suggesting strong acid rain with NOx and SOx.

The chemical compositions of filtered rainwater at the clinic and hotel in Cam Phuc showed quite high Sulfur content, with almost 50% S, suggesting that the NOx and SOx components of the pollutants originated from fossil fuels (Figure 3). The chemistry of the filtered rainwater agreed with the pH and EC data in Figure 1. The black materials on the beach sand in Cam Phuc and on the collapsed wall at the World heritage church building in Hue, were analyzed by ED-XRF, showing mainly Al, Si, K, Ca, and Fe associated S, Ti, and Pb. Note that heavy metals of Ti, Mn, Fe, Zn, and Pb were present in both black materials (Figure 3). Furthermore, NCS quantitative analyses of both black materials showed totals of 35 – 48 wt% of Nitrogen, Carbon, and Sulfur components of NOx, CO_2, and SOx, typical pollutants. Especially, the high Carbon content is remarkable, and it agrees with TEM observation results in Figure 2D.

3.2. Drinking Water Characteristics

Water samples were collected from wells and/or rainwater, ponds and irrigation canals in Cam Chinh and Cam Phuc and were measured for pH, Eh, EC, DO, water temperature and Rn in August, 2007 as shown in Table 1. The pH of the well water in Cam Chinh was acidic, whereas in Cam Phuc the pH of rain and well water indicated neutral pHs of 7, except for one of the wells, which had a pH of 4.4 (Table 1 shadow). The Eh, DO, and water temperature were almost the same between the 2 villages. The EC of most of the well water in Cam Phuc indicated 1-2 places higher (300-5000 μ S/cm) than that in Cam Chinh (several 10-100 μ S/cm). Especially the EC of the shrimp culture pond in Cam Phuc was 6 places higher (149 mS/cm) because of abundant feed. Fish with deformed backbones were found in the pond in Doc Kinh, Cam Chinh.

Both diagrams distinguished pH distribution between the two villages. The pH of the water in 30 wells in Cam Chinh clearly indicated strong acidic conditions such as pH4-5, whereas the pH of 32 samples were neutral at pH 7 in Cam Phuc (Figure 4). The pH difference between the two villages' drinking water is due to geochemical factors at depths of 10-20m. The reddish orange colored laterite soils at that depth were acidic in Cam Chinh village, whereas sandy or muddy sediments at relatively shallow depths were neutral in Cam Phuc. The two exceptions were wells of about 80m in depth in Cam Chinh which almost reached bedrock, and for that reason the water was no longer acidic, but pH6.5 (Figure 4).

The chemical analyses of well/rain water and coliform bacteria counts in Cam Chinh and Cam Phuc are shown in Table 2. Six well/ rain water samples showed counts as high as 100-202 colonies. The simplified water inspection products indicated high COD and NH_4^+ in 9 places (indicated by shading in Table 2). The XRF data showed chemical compositions in well/ rain water in both villages, indicating relatively higher contents of Na, Mg, S, K, and Ca elements in Cam Phuc, suggesting the effects of sea water in the shallow well. Note that in 2 places the Sulfur concentration was obviously high (25-38 wt%) in Cam Phuc. Some of heavy metals of Cr, Fe, and Zn were found in high concentrations in both villages. The Zn concentration of Doc Kinh clinic in Com Chinh is 2.2 mg/L which is twice as high as that of standard in Japan (1.1 mg/L). The B concentration of the river at 17°N latitude was as high as 1.6mg/L as compared to the standard in Japan (1.0 mg/L).

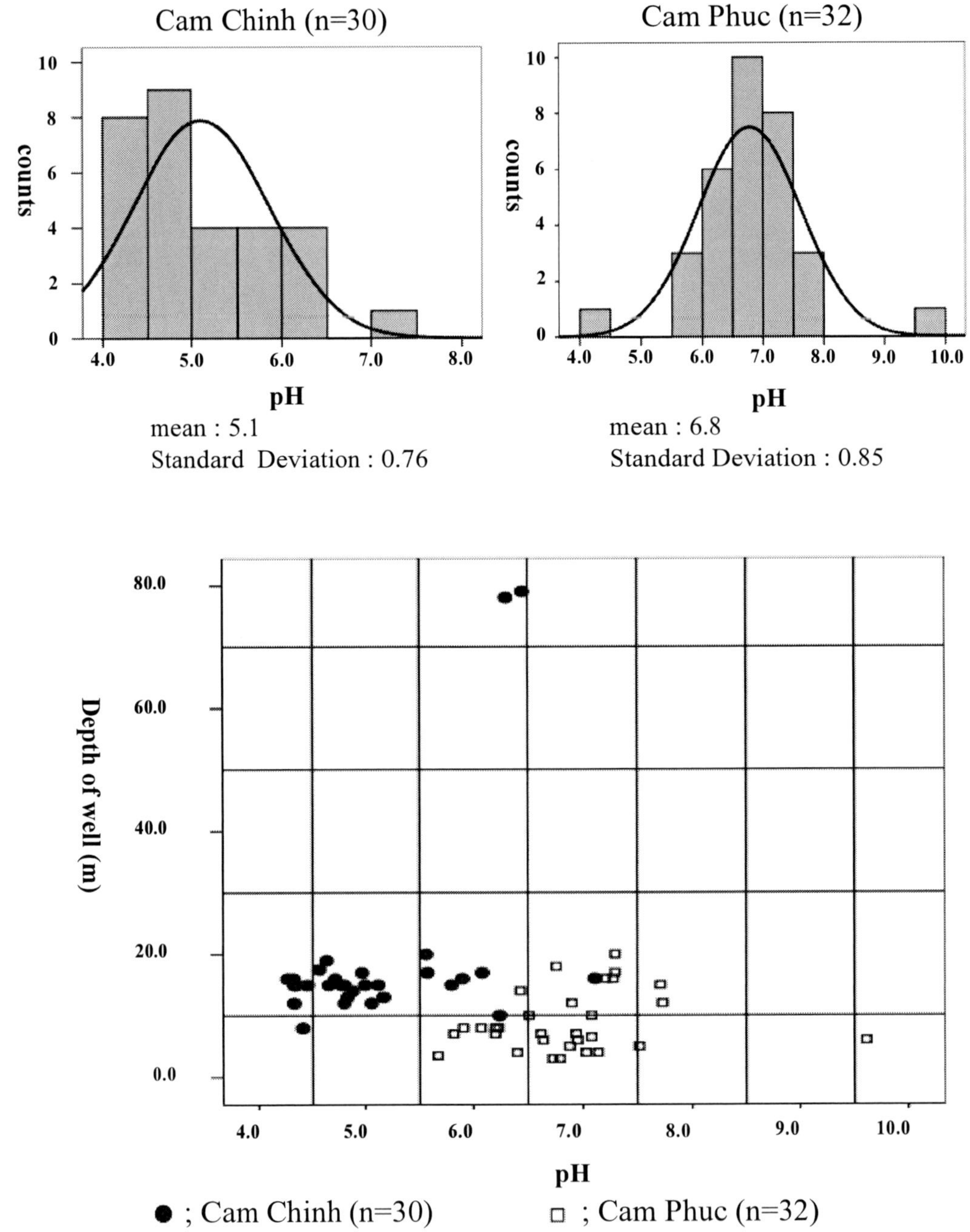

Figure 4. The pH of well-water at Cam Chinh and Cam Phuc, and the correlation of the depth and pH.

Table 1. Water quality in Cam Chinh and Cam Phuc (continued on next page)

Sampling points	settlement	water quality						water
		pH	Eh(mV)	EC(μS/cm)	DO (mg/l)	WT(℃)	Rn	
Cam Chinh	Loc An	4.7	497	38.1	6.3	24.9	85-90	well
	Mai Dam	6.5	455	108.9	7.4	25.9	75-85	PW
	Thiet Trang	4.8	518	35.5	6.7	25.0	65-70	well
	Thuong Nghia	4.8	503	34.1	7.4	29.8	60-65	well
	Mai Loc Ⅲ	4.3	532	33.9	4.6	25.2	100	well
	Mai Loc Ⅲ	4.3	541	31.3	6.2	25.2	60-65	well
	Trung Chi	4.8	388	40.6	5.4	25.8	55-60	well
	Mai Loc Ⅱ	5.8	464	43.7	6.4	25.7	70-80	pond
	Mai Loc Ⅰ	4.4	590	63.5	5.4	24.5	70-80	well
	Mai Loc Ⅰ	6.5	435	69.6	6.2	26.8	60	acequia
	Bien Dinh Do	4.6	591	48.6	5.3	25.0	70	well
	Doc Kinh	6.3	399	73.3	5.6	27.8	70	pond*
	Doc Kinh	5.1	504	37.2	5.6	25.6	60	well
	Doc Kinh	6.3	438	26.5	8.0	28.8	80	rice paddy
	Xom Moi	5.6	475	42.4	6.8	26.2	60	well
	Phuong An	4.6	568	51.2	6.4	26.1	60	well
	Doc Kinh	6.3	536	107.9	6.8	25.1	90-100	well
	Minh Huong	5.1	444	113.0	4.3	24.8	80-100	well
	Mai Loc Ⅱ	5.9	616	48.7	7.5	25.0	80	well
	Son Nam	6.1	483	194.5	4.9	27.0	50-60	well
	Doc Kinh clinic	6.2	463	186.0	7.6	26.8	80-90	well

Table 1. Continued

Sampling points	settlement	water quality						water
		pH	Eh(mV)	EC(μS/cm)	DO (mg/l)	WT(℃)	Rn	
Cam Phuc	Cam Phuc clinic	7.5	476	454.0	11.4	29.2	-	well
	Thou 5	6.6	445	1187.0	5.6	29.0	-	well
	Thou 5 junior high	8.9	444	64.5	5.9	27.3	80	rain
	Thou 6	7.7	411	64.0	4.9	29.1	-	rain
	Thou 8	7.2	393	63.0	4.0	28.5	-	rain
	Thou 9	8.3	260	89.6	6.9	28.0	-	rain
	Thou 13	6.7	415	210.0	6.1	30.2	60	PW and R
	Thou 15	6.6	91	5000.0	2.3	26.8	90-100	well
	Thou 15	6.5	80	4900.0	1.8	25.8	100	well
	Thou 12	6.4	359	2080.0	3.3	26.8	60	well
	Thou 11	4.4	607	288.0	3.4	27.6	70-80	well
	Thou 10	6.2	179	375.0	3.9	29.2	100-110	R and W
	Thou 3	6.8	105	1830.0	3.8	27.0	70-80	well
	Thou 2	7.1	401	1203.0	3.9	28.0	-	well
	Thou 1	6.9	135	322.0	2.0	28.0	-	well
	Thou 4	5.9	214	399.0	3.1	28.0	80	well
	Thou 7	7.3	54	2400.0	2.1	27.0	80-90	well
	Thou 14	7.9	340	91.7	3.7	27.5	-	rain
	Thou 14	7.6	383	2.5	4.9	29.0	-	rain
	Thou 14	7.9	337	149000.0	5.5	30.5	-	shrinp pool

well : well water
rain : rain water
PW : public well water
R and W : mixed rain water and well water
PW and R : public well water and rain water
pond* : deformed backbone fishes are found in this pond.

3.3. Family Structure

The ages, gender distribution, and family structure of the 28 and 30 participants for health check interviews at Cam Chinh and Cam Phuc villages respectively were shown in Figure 5. Mostly females between 40-50 years old were interviewed for the health check questionnaire. Most families live with 2-3 generations in the same house, but the ratio of 2 and 3 generation households in both villages is different. Most families had 4-5 members and families of more than 6 are common in Cam Phuc, meaning many children, which is an indication of a poor economic situation. The results suggest that not only the low quality of drinking water, but also many children, lower income, and poor nutrition effect their health conditions.

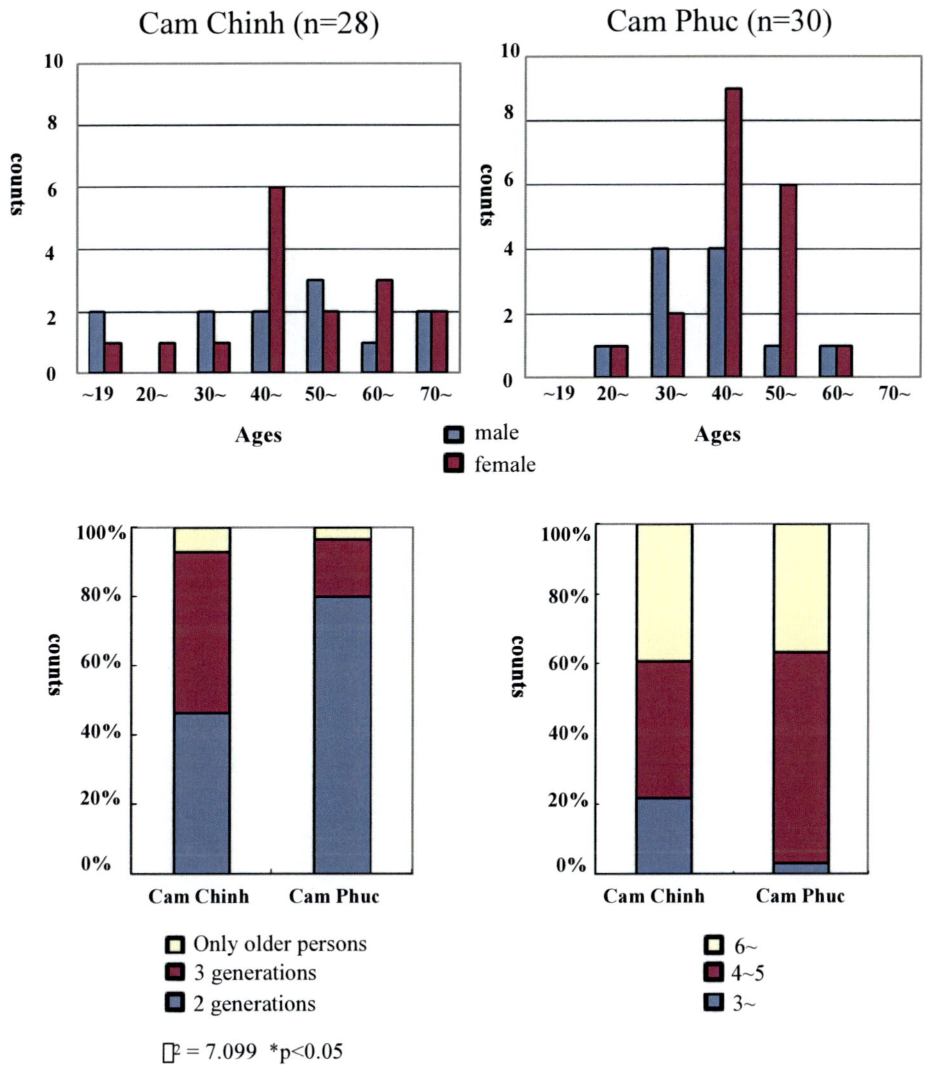

□^{2} ; square test p ; significant difference

Figure 5. The age, gender distribution and family structure of the participants in Cam Chinh and Cam Phuc.

3.4. Human Health

The quality of drinking water is related to health conditions, which differ between Cam Chinh and Cam Phuc (Figure 6). The answer of "sometimes" and "always" for fatigue, loss of appetite, and stomachache were higher in Cam Phuc than in Cam Chinh, but more people reported eating uncooked foods in Cam Chinh than in Cam Phuc. The results of the questionnaire suggest that cooked foods have been kept their good health conditions, even though the drinking water was strong acidic conditions.

The check data of anemia, roundworm, dermatitis, and dizzy factors between two villages, indicated the same tendency as feeling tired, less appetite, and stomachache factors, that are not in graphed here.

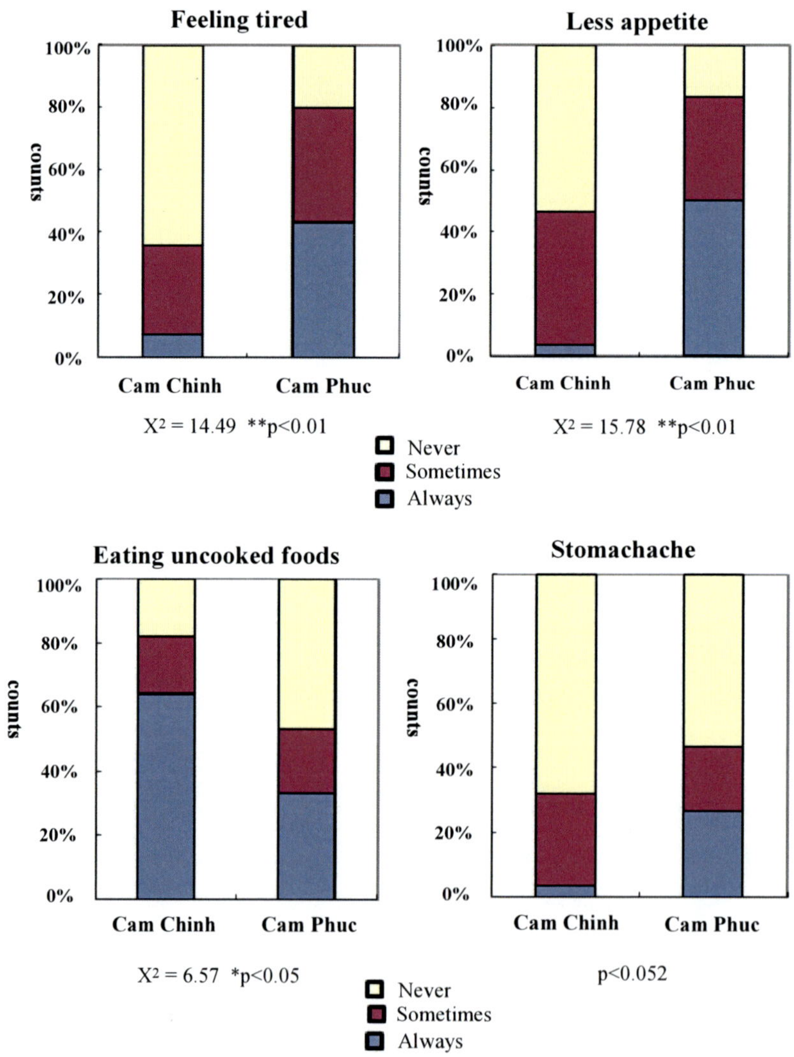

□2 ; square test p ; significant difference

Figure 6. The health check with interviews at Cam Chinh and Cam Phuc.

3.5. Dinking Water Maintenance of Well and Rain Water

In Cam Chinh, 96% of drinking well water is boiled, whereas 63% of the rainwater is boiled and the other 37% of well and/or rain water are filtered and/or boiled in Cam Phuc Village. The type of filtration methods and layered materials in the jar container were varied in Cam Phuc Village, such as fine sand only (2), fine sand + pebbles (1), fine sand + pebbles +charcoal (1), fine sand + charcoal (3), fine sand +charcoal + pebbles + coarse sand (2), and fine sand + pebbles + charcoal + coarse sand (2), suggesting that purification methods are different from each other. In Cam Phuc 37.5% of the wells were covered with 62.5% of wells being open, whereas in Cam Chinh 96.7% of the wells were covered. Serious problems were found with the filtration systems in Com Phuc, namely the period of cleaning the containers. The filtration jar in the best condition was found at the high school in Cam Phuc, where one a month periodical cleaning was reported. On the other hand, in ordinary houses the jars were cleaned only once every half year to 3 years. That allowed for the occurrence of mosquito larva and bio-films in the jars, contaminating the drinking water.

3.6. Volatile Organic Compounds in the Air

Volatile organic compounds were measured on August 2007 at 5 places in Cam Chinh, Ha Tinh, Hue, and Com Phuc. The concentrations of several elements are clearly higher than the standard in Japan, such as toluene, hexane, formaldehyde, acetaldehyde, acetone, propionaldehyde, hexaldehyde, and 2,5-dimethylbenzaldehyde (Table 3. shadows). Especially acetone concentrations at all of the 5 places were unbelievably 100 times higher than the standard in Japan, but the origin of the acetone is unknown.

4. Discussion

4.1. Dioxin Contamination

Between 1962 and 1971 in Vietnam War, the US military sprayed herbicides mainly in the south part of Vietnam; the operation was named Ranch Hand. Organic chlorine herbicides including Agent Orange were contaminated with dioxin, which is a kind of endocrine disrupting chemical (EDC). In 1974, the US National Academy of Sciences published estimates of the extent and distribution of herbicides sprayed, but a recent study by Stellman et al., (2003) reported that the spray inventory with heavily dioxin-contaminated herbicides is estimated to be seven million liters more and that the amount of dioxin sprayed is estimated to be almost doubled. Mean concentrations of PCDDs, PCDFs, and Total-PCDD/Fs in adipose tissues of inhabitants in the herbicide-contaminated area were significantly higher than those in the non-contaminated area. However, no reasonable associations were found between Total-PCDD/Fs and some causal factors such as age, BMI, military records or contact with pesticides (Kido et al., 2007). In this study dioxins were not found in environmental samples collected from central Vietnam in 2006 and 2007.

4.2. Large Scale Transfer and Long Distance Travel of Air Pollutants

Measurements of midday vertical atmospheric CO_2 distributions reveal annual-mean vertical CO_2 gradients that are inconsistent with atmospheric models that estimate a large transfer of terrestrial carbon from tropical to northern latitudes. This suggests that northern terrestrial uptake of industrial CO_2 emissions plays a smaller role than previously thought and that, after subtracting land-use emissions, tropical ecosystems may currently be strong sinks for CO_2 (Stephens et al., 2007). Uno (2003) reported a large transfer of terrestrial black carbon particles from tropical latitudes such as South West Asian countries. In the black carbon hazard map, black carbon particles are concentrated 0.1-0.2g/m^2 in Vietnam, Laos, Thailand, Cambodia, Myanmar, Bangladesh and India, suggesting carbon soot inflow to Vietnam from Western countries. Recently, the production of NOx in Asian countries has tripled the past 20 years. Especially, China and India use oil and coal energies for power plants, factories, and automobiles, therefore the NOx is increasing dramatically. Emissions from the world's 2100 coal fired power plants are responsible for roughly a third of the CO_2 generated by human activity. Power plant, easily surpass the amount produced by cars and all other industries combined. Additionally, the share of electricity generated by coal in India is expected to climb year by year. Last year, China, which delivers about 80% of its electricity from coal and recently surpassed the United States as the world's biggest CO_2 emitter, brought online two major coal plants a week.

Furthermore, Tazaki et al., (2004a, b) reported the transport of carbon-bearing dusts from Iraq to Japan during the Iraq War, finding abnormal events during the Iraq War monitored at Japan 8000 km away. Tazaki and Asada (2007) reported transmission electron microscopic observation of mercury bearing bacterial clay minerals from a small-scale gold mine near Lake Victoria transported to Dar Es Salaam more than 1000 km away in Tanzania.

In this study, the heavy metals and black carbon particle contamination of rainwater in Vietnam may be, if anything, negligibly small on undetectable levels in the summer wet season in Vietnam. But, the powdery dusts and carbon-bearing particles produced by combustion of oil and gas in South West Asian countries are directly transferred to Vietnam by Westerly winds during the winter dry season (Uno et al., 2003). Black carbon particles are products of incomplete combustion, and are the principal light-absorbing atmospheric aerosols as large-scale transfer and long distance air pollutants. In the study sites, neither industrial areas nor combustion sources were found in Vietnam.

4.3. Air Pollution by Formaldehyde (FA)

Indoor air quality has been affected by a variety of chemicals, which are suspected to be the cause of sick building syndromes around our living conditions. Tazaki et al, (2002) reported that environmental survey of indoor Air pollution by Formaldehyde (FA) in Kanazawa, Ishikawa, Japan. The 84 locations were selected from various living environments such as homes, school, bookstores, and cars in Kanazawa city. FA from 51/84 locations (60.7%) detected more than 0.02ppm whereas FA concentrations from 25/84 locations (29.8%) detected more than 0.04ppm, indicated FA on low-level and long-term basis. The guideline of FA concentration recommended by the Ministry of Health, Labour and Welfare Japan is 0.08ppm. FA concentration increased with time according to the rise in room temperature, was emitted from building materials, such as wall, ceiling, floor materials and adhesive agents. The life styles associated with heating –system, furniture and cigarette smoke, are also responsible for the emission of FA.

In this study, volatile organic compounds at 5 places in Cam Chinh, Ha tinh, Hue, and Com Phuc are clearly higher than the standard in Japan, such as toluene, hexane, formaldehyde, acetaldehyde, acetone, propionaldehyde, hexaldehyde, and 2,5-dimethylbenzaldehyde. Especially acetone concentrations are unbelievable 100 times higher than the standard in Japan which is unknown the origin of acetone.

4.4. Heavy Metal Pollution of Drinking Water

Underground water has been extensively used as drinking water for the last 20-30 years for the reduction of infant mortality from diarrheal diseases in Bangladesh. The tube-well water is polluted with high concentrations of arsenic and poisoning has become a serious public health problem in Bangladesh. In geo and aquatic ecosystems microbial mats of bacterial colonies play an important role in the remediation of toxic heavy metals in polluted soil and water systems. Investigations were carried out to qualify the arsenic accumulation by microbial mats for bioremediation (Islam and Tazaki, 2003). But, very little information about heavy metals in drinking water in Vietnam has been reported. In this study, we found heavy metal ions of Mn, Fe, Sr, and Cu in well and filtered rainwater in the villages.

4.5. Soils in Vietnam

Well water related geology and soil mineralogy has produced characteristic water quality. Rainwater included deposits such as fine grain sediments and dusty fossil fuel materials. We do see evidence of strong acidic or neutral drinking water modified by geologically recent activity in Vietnam. Predominant clay minerals in the Red River Delta, northern Vietnam, are mica and kaolinite; the mica and vermiculite are higher for alluvial soils while the kaolinite and chlorite-vermiculite intergraded minerals are higher for grey degraded soils. The lower mica and higher kaolinite contents in the grey degraded soils can be considered as an indication of the early stage of soil degradation (Hai et al., 2006).

Sixteen profiles of alluvial soils of Vietnam were collected throughout the country and subjected to clay mineral analysis (Hai and Egashira, 2005). The clay mineralogical composition was remarkably similar within soil units belonging to the same river system and in a profile. However, the clay mineralogical composition greatly varied with river systems; therefore, regional variation existed and was mainly controlled by the mineralogical features of the sediment sources.

In this study, the main clay mineralogical components in Cam Chinh and Cam Phuc Villages, were kaolin minerals in laterite soils. The wells 20-30m in depth in both villages need to be dug deeper to go through the laterite soils. Otherwise they need to clean the filter systems more often to keep clean, safe drinking water. There were different filtration methods with or without covers for wells in the villages, and from place to place, so that is a variable that could have affected the results of the study. Furthermore, we must find new sustainable filtering, purification, and storage methods associated with their effectiveness, which are safe, less costly, easy to use and maintain, and which use local natural materials. They must void secondary effects by any human made chemical materials.

4.6. Environmental Education and Social Sciences

The relationships between low quality drinking water, uncooked foods, wells without lids, number of children per family and unhealthy conditions are totally related to life conditions in the villages. Moreover, low income and poor nutrition affect their health conditions. Both environmental Earth Science and Social Science education are also needed in the countryside. They do not need high technology or expensive imported filters at all; they just need to clean the filter systems more often, and store clean safe water in covered containers.

5. Conclusions

Historical monuments, statues, and buildings in Vietnam are covered by black soot and fly ash caused by acid rain. The sooty surfaces are also damaged by fungi, lichens, and bacteria. Electron microscopy observation showed abundant carbon spherules and fly ashes, associated with microorganisms in rain water. Most Vietnamese drink water from shallow wells and rainwater, especially in the countryside. Sometimes the rainwater showed strong acidic pH and high Electrical Conductivity during dry season, suggesting high contents of heavy metal ions associated with SOx, NOx, and CO_2. The relationships between low quality drinking water, uncooked foods, wells without lids, and number of children per family are totally related to poor health conditions and shorter life expectancy. They need to clean the filter systems more often, and keep the clean safe water covered. Furthermore we must find new filtering systems, which are safe, low cost, easy, and sustainable, using local natural materials without manmade chemical substances. It is time to act locally in Vietnam and think globally about safe drinking water.

Acknowledgments

We would like to thank the scientists of Nhu Tren and Luu Ban 10-80 groups, interpreters, and drivers in Vietnam who helped us in our field works. We also express our sincere thanks to Keely Brandon for checking and improving the English. This study was supported by grants from the Japanese Ministry of Education, Science and Culture.

References

[1] Cho, A. (2007). Forecasting global warming's monumental impact. *Science*, **316**, 1547.

[2] Dung, P.T., Hung, T.M., Kido, T., Tawara, K., Thom, L.T.H., Nhu, D.D., Naganuma, R. (2007). The effects of agent orange/dioxin on pregnancy outcomes in a sprayed area in Viet Nam. *Organohalogen Compounds*, **69**, 564-567.

[3] Hai, N.Q. and Egashira K. (2005). Clay mineralogical composition of alluvial soils from different river systems/agro-ecological regions in Vietnamin reference to origin, regional distribution and soil quality. *Clay Science*, **12**, 349-360.

[4] Hai, N.Q., Tra, H.T.L., Moslehuddin, Z.M. and Egashira K. (2006). Clay mineralogy of Hlluvial soils and grey degraded soils derived from the same river sediment in the Red River Delta, Northern Vietnam. *Clay Science*, **13**,101-105.

[5] Islam A. R., Tazaki K. (2003). Bioremediation of As polluted groundwater in Bangladesh. In: Tazaki K., ed., *Water and Soil Environments*: Microorganisms play and important role: Kanazawa University 21st-Century COE Program, p. 25-52.

[6] Kido, T., Maruzeni, S., Suzuki, H., Odamae, Y., Muranaka, M., Naganuma, R., Tawara, K., Nishijo, M., Nakagawa, H., Hung, T.M., Thom, L.T.H., Dung, P.T., Nhu, D.D., Oka, H., Noguchi, K. (2007). Five years studies on the long term effects of war agent orange /dioxin on human health in Vietnam. *Organohalogen Compounds*, **69**, 572-575.

[7] Nishijo, M., Tawara, K., Hung, T.M., Thom, L.T.H., Dung, P.T., Nhu, D.D. (2007). An epidemiological study on health effects by dioxin in Vietnam; Comparison of visual acuity of junior high school students between herbicide sprayed and non-sprayed areas. *Organohalogen Compounds*, **69**, 2035-2038.

[8] Sawano, N., Nhu, D. D, Kido, T., Tawara, K. (2007). Geo-spatial approach to herbicide spraying during Vietnam War. *Organohalogen Compounds*, **69**, 560-563.

[9] Stellman, J.M., Stellman, S.D., Christian, R. et al. (2003) The extent and patterns of usage of Agent Ornge and other herbicides in Vietnam. *Nature*, **422**, 681-687.

[10] Stephens, B.B., Gurney, K.R., Tans, P.P. others 13 (2007). Weak Northern and Strong Tropical land Carbon uptake from Vertical profiles of Atmospheric CO_2. *Science*, **316**, 1732-1734.

[11] Tawara, K., Nishijo, M., Nakagawa, H., Kido, T., Naganuma, R., Suzuki, H., Hung, T.M., Thom, L.T.H., Dung, P.T., Nhu, D.D. (2007). Association of having birth defect children with serum dioxin levels in Vietnam. *Compounds*, **69**, 568-571.

[12] Tawara, K., Nishijo, M., Honda, R., Nakagawa, H., Seto, T., Kido, T. (2007). Relationship between food preference in women and PCDDS/DFS levels in breast milk in the northern district of Japan. Organohalogen *Compounds*, **69**, 2031-2034.

[13] Tazaki, K. and Asada, R. (2007). Transmission electron microscopic observation of mercury bearing bacterial clay minerals in a small-scale gold mine in Tanzania. *Geomicrobiology Special Issue on Geomicrobiology Processes in Extreame Environments*. Ed: Hailing Dong and Chuanlun Zhang. 24, 477-489.

[14] Tazaki K., Wakimoto R., Minami Y., Yamamoto M., Miyata K., Sato K., Saji I., Chaerun S. K., Zhou G. P., Morishita T., Asada R., Segawa H., Imanishi H., Kato R., Otani Y., Watanabe H. (2004a). Transport of carbon-bearing dusts from Iraq to Japan during Iraq's War. *Atmospheric Environment*, v. 38, p. 2091-2109.

[15] Tazaki K., Miyata K., Tanaka T. Y., Chiba T., Wakimoto R., Saji I., Sato K. (2004b). Finding abnormal events during the Iraq War monitored at Japan 8000 km away form Iraq. *Science Reports of Kanazawa University*, v. 48, p. 107-133.

[16] Tazaki, K., Segawa, S., Minami, M., Moriichi, S. (2002). Environmental survey of indoor Air pollution by Formaldehyde (FA) in Kanazawa, Ishikawa, Japan. *Earosol Research*, **17**, 284-290 (in Japanese with English Abstract).

[17] Thom, L.T.H., Hung, T.M., Kido, T., Tawara, K. (2007). Agent Orange health related outcomes of population living in herbicides sprayed area of Quang Tri Viet Nam by epidemiological cohort study. *Organohalogen Compounds*, **69**, 2148-2151.

[18] Uno, L. (2003). Regional chemical weather forecasting system CFORS: Nodel descriptions and analysis of surface observations at Japanese Island stations during the ACE-Asia experiment. *J. Geophys. Res.*, **108** (D23), 8668, doi: 10.1029/2002JD002845.

In: Acid Rain Research Focus
Editor: Lyman G. Roglesfield, pp. 143-159
ISBN: 978-1-60456-373-3

Chapter 6

SORPTION OF SULFATES, RETENTION OF CATIONS AND ACID RAIN BUFFERING CAPACITIES OF FOREST SOILS IN CENTRAL TAIWAN

Ming Kuang Wang[1], Shun Yao Zhuang[1,2], Po Neng Chiang[1] and Hen Biau King[3]

[1]Department of Agricultural Chemistry, National Taiwan University, Taipei, Taiwan, 106
[2]State Key Laboratory of Soil and Sustainable Agriculture, Institute of Soil Science, Chinese Academy of Science, Nanjing, 210008, China
[3]Taiwan Forestry Research Institute, Council of Agriculture, Taipei, Taiwan, 106

Abstract

The subtropical alpine forest ecosystem is vulnerable to acid rain deposited through the forest soils. The average sulfate and nitrate deposited by acid rain as precipitation amount to 50 kg SO_4^{2-} and 15-50 kg NO_3^- ha^{-1} year in Taiwan, respectively. Soil samples were leached with simulated acid rain solutions of various acidity and sulfate concentrations to evaluate their retention of cations and sulfates, testing and leaching with break through curves with simulated acid rain solutions. The pH buffer capacity of forest soils along an elevation sequence in central Taiwan was also investigate through conventional titration and mathematical calculation of proton affinity methods. The amounts of sorbed-SO_4^{2-} of two pedons extracted with NaH_2PO_4 (0.016 M) solutions in the range 0.14-1.09 mmol kg^{-1} soil were higher than that of native water-soluble sulfate (i.e., extracted with H_2O), in the range 0.15-0.41 mmol kg^{-1} soil. The sorption of sulfate did not fit well the Langmuir equation. The sulfate sorption capacity and the amount of released Al correlated well with increasing acidity of simulated acid rain added. Released K, Mg and Ca showed two periods of higher leaching pattern, having a trend similar to that the electric conductivity (EC) pattern. The pH buffer capacity (pHBCs) showed no significant differences from the titration curves of various pedons and horizons. However, mathematical calculation of proton affinity potentials showed the highest correlation at the lower elevation site with soil pH ($r = 0.90^{**}$), cation-exchange capacity (CEC) ($r = 0.91^{**}$), base saturation ($r = 0.97^{**}$), and X-ray non-crystalline sesquioxides ($r = 0.68^{**}$). It is important to evaluate the alpine forest soil pH buffer capacity by combining its proton consumption potentials.

Keywords: Acid-buffering capacity, acid deposition, cations, leaching column, soil buffering capacity, soil proton affinity.

1. Introduction

The chemistry of acid rain deposition in Taiwan was first reported in 1980, but the investigation included no systematic monitoring. The monsoon over the island was acidic and concentrated during May and June every year. The average pH was around 4.80 between 1980 and 1986 [1]. A systematic monitoring network with 10 sites was established to investigate the chemistry of precipitation on the island since 1990. The average depositions of sulfate and nitrate from acid rain were 50 kg SO_4^{2-} ha^{-1} $year^{-1}$ and 15-50 kg NO_3^- ha^{-1} $year^{-1}$, with precipitation rates of 23-57 SO_4^{2-} mg $(m^2)^{-1}$ s^{-1} and 24-170 NO_3^- mg $(m^2)^{-1}$ s^{-1}, respectively [2].

The formation of acid rain is primarily due to SO_4^{2-} emission from the burning of fossil fuels by industries, as well as NOx effluent from mobile vehicles and industrial boilers. The emitted SO_4^{2}- and NOx are then transformed photochemically to become sulfuric and nitric acids, and transported over long distances before falling to the ground as components of precipitation, causing damage to our ecological system. The harmful effects of acid rain include corrosion of materials and structures, dieback of forests, mobility of sulfate in soil profile, as well as acidification of lakes and soils [3]. High amounts of sulfate and nitrate are deposited on the island as a result of industrial pollution [4,5]. Few investigations on the interaction of acid deposition with forest soils in the tropics and subtropics have been made. The subtropic montane forest ecosystem is vulnerable to acid leaching through the forest soils [6-10].

Soil contains different amounts of materials that possess pH buffer capacity [11], acting to prevent abrupt alteration of pH on addition of acid or base [12]. Soil sesquioxides act as sorbents to sorbed SO_4^{2-} and NO_3^- compounds from acid deposition [13]. Some compounds in soils have a variable charge that increases positive charge at low pH; these soils with a large cation-exchange capacity (CEC) from organic sources also lose this CEC at an acidic pH as carboxylate groups are prontonated or binded with aluminum ions [14]. Soil potential in pH buffer capacity could be assessed through measuring simultaneously the soil sulfate retention and pH capacity. The aims of this report include two sections of (1) sorption of sulfate and retention of cations in forest soils, and (2) rain buffer capacities of alpine forest soils in central Taiwan.

2. Environmental Conditions and Soil Sampling of Study Sites

1. Two pedons were collected from the Lien-Hua-Chi watershed region in central Taiwan. The sample sites are located at latitude 23°33'N and longitude 120°23'E [9] with elevations of 723 m and average slope of 69%. The climate is humid temperature with mean annual temperature of 20.6°C and mean annual precipitation of 2200 mm, 80% of which occurs between May and September [15]. The natural vegetation consists of evergreens mixed with hardwoods, typical of warm temperate montane rain forest [16]. The bed rock consists of interbedded Tertiary sandstones

and shales [17,18]. Two soil samples were obtained from the surface and subsoils. Site 1 represents a fine mixed, thermic, Typical Dystrochrept (Inceptisols). Site 2 is sandy clay, mixed thermic, Typical Hapludult (Ultisols) [19]. Pedon 1 was divided into four horizons, namely 0-10 (A1), 10-30 (A2), 30-50 (Bw) and 50-80 cm (BC), and pedon 2, three horizons of 0-10 (A), 10-40 (E), and 40-60 cm (E/B).

2. A-Li-Shan region is one of the famous montane sceneries in Taiwan. The environmental conditions are similar to that of the Lien-Hua-Chi watershed region. There were four pedons (A-D) collected along an elevation sequence at this region. The westward facing slopes of the selected sample sites were once the sources of acid deposition. The sample sites are located (i) pedon A (Lu-Lin Shan, LLS, elevation 2700 m, vegetation of *Pinus taiwanensis*), at latitude 23°27'N and longitude of 120°51'E; (ii) pedon B (Ta-Ta-Chia, TTC, 2620 m in elevation, *Pinus taiwanensis*), at latitude 22°29'N and longitude of 120°53'E; (iii) pedon C (A-Li Shan, ALS, 2350 m in elevation, *Cryptomeria japonica*), at latitude 23°31'N and longitude of 120°44'E; (iv) pedon D (Shih-Tzu-LU, STL, 1500 in elevation, *Cryptomeria japonica*), at latitude 23°29'N and longitude of 120°45'E.

3. Soil Analysis

Soil samples were dried in air and crushed to pass through a sieve (2 mm). The pH of soil was determined in a mixture of soil and water at a 1:10 ratio and in a 0.01 M KCl suspension. Organic C was determined with the wet oxidation method of Walkley and Black [20]. Cation-exchange capacity (CEC) was determined by conventional $Na-NH_4$ exchange method [21]. The distribution of particle size was determined with hydrometer [22]. The silt was separated from the sand by wet sieving (sieve 53 μm) method.

Free iron and aluminum oxides were extracted with dithionite, citrate and bicarbonate (Fed, Ald) [23]. X-ray noncryatalline iron and aluminum were extracted with ammonium oxalate (0.2 M, pH 3) (Feo, Alo), and organic-bound contents were extracted with Na-pyrophosphate solutions (0.1 M, Fep, Alp) [24]. Exchangeable cations were extracted with ammonium acetate solution (1 M). The Lien-Hua-Chi soils were subjected to do the following tests:

3.1. Sulfate Sorbed of Soils

Native sorbed sulfate of soils were divided into soluble and adsorbed sulfate. 3 g of soils were extracted with deionized water (30 mL) to determine the water-soluble sulfate. Soils (3 g) were equilibrated with a NaH_2PO_4 solution (0.016 M, 30 mL) as sorbed sulfate, and centrifuged at 21,400 xg for 20 min. All tested were performed in triplicates. Sulfate concentrations in the clear supernatant were determined with an ion chromatograph (Dionex-2000i, IC). Total native sorbed sulfate of soils was the sum of both water-soluble and desorbed sulfate.

3.2. Sulfate Sorption Isotherm and Kinetic of Sulfate Sorption

3 g of soil samples were placed into a polyethylene tube (50 mL) equilibrated with Na_2SO_4 solutions (0.01, 0.02, 0.05, 0.1, 0.2, 0.5 and 1.0 mM, 30 mL), shaken for 2 h and then centrifuged to determine sulfate sorption. For the kinetic study,).1 mM, 30 mL of Na_2SO_4 solutions at 25°C, shaken at 150 rpm for 0.5, 1, 2, 4, 8, 16 and 24 h, and then centrifuged. Sulfate concentrations in the clear suspernatant were determined with IC.

3.3. Effect of pH on Sulfate Sorption and Released of Cations

To determine sulfate sorption in batch experiments, 3 g of soils were placed into a polyethylene tube (50 mL). The Na_2SO_4 concentrations were 0.01, 0.1 and 1.0 mM, and the pH was adjusted to 2.6, 3.6, 4.6 and 5.6 with HCl and NaOH solutions. After equilibration on the shaker for 2 h, the suspension was centrifuged and the sulfate concentration in the clear supernatant was determined with an IC; the concentrations of released cations were determined with an atomic absorption spectrometry (AAS, Hitachi 180-30) and flame photometry.

3.4. Soil Columns Leached with Simulated Acid Rain

In order to assess the influence of acidity on retention of cations and sulfate, forest soils were artificially leached with simulated acid rain. The composition of the simulated acid rain at various pH is listed in Table 1. Variation in pH was achieved by adding various amounts of H_2SO_4 solutions.

An acrylic column of soil of 10 cm length (internal diameter of 2.54 cm, soil particle size less than 2 mm) was collected from each soil horizon of pedons 1 and 2 (Lien-Hua-Chi watershed region), which a simulated acid rain solutions was introduced. The chemical compositions of simulated acid rain solutions of the leaching columns are listed in Table 1. The bulk density of soils in a column was 1.30 g cm^{-3}. The flow rate was adjusted with a mechanical vacuum extractor. Flow velocity observed from average saturated hydraulic conductivity of soils was 1.0 cm h^{-1}. Pore volume of solution was calculated from saturated water hydraulic conductivity, porosity and total volume of soils in column. Percolate solutions were collected and analyzed for sulfate and cation concentrations. The amounts of sulfate retained by soils were calculated from the difference between the amounts of sulfate present in the leaching solutions and the leachates. The EC was determined with a microprocessor conductivity meter (Orion model 120).

Soils from the A-Li-Shan region were conducted the following experiments:

Table 1. Composition of simulated acid rain

Simulated acid rain (number)	pH	SO_4^{2-} (mmol L^{-1})	EC (dS m^{-1})
1	2.6	1.491	905.0
2	3.6	0.110	74.8
3	4.6	0.009	10.9

3.5. Simulation Analysis of Soil pH Buffer Capacity

In the experiment, 30 mL of simulated acid rain solutions was added into a 50 mL centrifuge tube containing 2 g of soil. All tubes were shaken at 150 rpm under 25±0.2°C for 24 h and then centrifuged under 21,400 xg for 10 min. Supernatant was collected and pH was measured simultaneously. The pH buffer capacity (pHBC) was calculated as the negative slope from the amount of acid added (*X* axis) and pH (*Y* axis) [10,25].

3.6. Proton Affinity Potential Calculation

The proton consumed by soil ($H^+_{consumed}$) was calculated using the following equation [26]:

$$H^+_{consumed}\,(\mathrm{cmol\ kg^{-1}}) = \{Vo(C_a\text{-}C_b) + \Delta V/N_t\text{-}(Vo + \Delta V)([H^+]_f - [OH^-]_f)/Wo \qquad (1)$$

Where C_a and C_b are analytical concentrations of acid (or base) added initially, $[H^+]_f$ and $[OH^-]_f$ are actually measured concentrations, Vo is the volume of solution corresponding to the starting point of titration, ΔV is the cumulative volume of titrant added, and N_t is the normality of titrant (negative for titration with base). The proton and hydroxyl concentrations and their activity coefficient (Υ_i) can be calculated using David' equation:

$$\mathrm{Log}\,\Upsilon_i = -\,0.5115Z^2(I^{1/2})/(1 + I^{1/2}) - 0.3I \qquad (2)$$

where I is the ionic strength for each point of the titration. The normalized proton consumption function was defined with respect to the amount of titrated soil samples. The proton affinity was described by the slope of proton consumption in soils (cmol kg^{-1}) versus solution proton concentrations (mM).

4. Results and Discussion

4.1. Soil Chemical and Physical Properties

The texture of two pedons (Lieh-Hua-Chi watershed region) ranged from clay to sandy clay loam (Table 2A). The mean values of pH were in the range 4.20-4.66 (in H_2O), and 3.33-3.65 (in KCl). The mean values of organic C in each pedon generally decreased with increasing soil depth. The range of CEC was from 10.93 to 19.42 cmolc kg^{-1} soil. The amount of exchangeable cations indicated that high exchangeable Al was present in all pedons. The highest soil pH of 5.3-5.3 was found in the pedon of Shih-Tzu-Liu (STL), while the lowest soil pH 3.7-4.2 was found in the pedon of A-Li Shan (ALS) (Table 2B). The organic carbon in the A horizon of ALS pedon was 98.8 g kg^{-1} and showed the highest among all soil samples. The exchangeable acidity carried with different pedons. Cation-exchange capacity (CEC) was the highest in the pedon of STL among all soil pedosns and it was the lowest in Lu-Lin Shan (LLS) pedons. Similarly, base saturation was the highest in the STL pedons and the lowest in the LLS pedon. Soil texture was even in the pedon of STL as sandy loam and varied slightly in other pedons.

Table 2. Chemical and physical properties of soil tested (A) Lien-Hua-Chi and (B) A-Li-Shan regions

(A) Lien-Hua-Chi watershed region

Soil	Horizon	Depth	pH (1:10)		Particle size (g kg^{-1})			Texture	Organic carbon	Exchangeable cations (mmol kg^{-1})					CEC (cmol kg^{-1})	BS	Extractable SO_4^{2-} (mmol kg^{-1})	
		(cm)	H_2O	KCl	Sand	Silt	Clay		(g kg^{-1})	K	Na	Ca	Mg	Al		(%)	H_2O	NaH_2PO_4
Dystrochrept (Pedon 1)	A1	0-10	4.31	3.54	71	406	523	C	21.8	2.98	0.05	0.11	0.68	5.50	18.11	2.1	0.41	0.92
	A2	10-30	4.56	3.64	65	358	577	C	13.5	2.29	0.06	0.04	0.37	5.94	17.49	1.6	0.38	1.09
	Bw	30-50	4.63	3.65	68	357	575	C	7.0	1.83	0.23	0.01	0.30	8.10	18.05	1.3	0.23	0.64
	BC	50-80	4.66	3.63	8	354	638	C	4.6	1.95	0.29	0.03	0.21	9.16	19.42	1.3	0.15	0.14
Hapludult (Pedon 2)	A	0-10	4.20	3.33	493	250	257	SCL	21.6	0.96	0.01	0.07	0.19	7.43	12.54	1.0	0.24	0.31
	E	10-40	4.44	3.56	468	245	287	SCL	4.1	0.55	0.11	0.04	0.09	5.83	10.93	0.7	0.32	0.73
	E/B	40-60	4.44	3.54	504	210	286	SCL	4.1	0.54	0.02	0.02	0.06	5.86	11.46	0.6	0.32	0.85

Table 2. Continued

(B) A-Li-Shan region

Soil Pedon	Horizon	Depth (cm)	pH	Org-C ($g\ kg^{-1}$)	Exchangeable Acidity	K	Exchangeable Na	Ca	Mg	CEC	Base Saturation (%)	Texture
					---------------	---------------	$cmol(+)kg^{-1}$	---------------	---------------			
Pedon of Lu-Lin-Shan (LLS): 2720 m												
1	A	0-10	4.3	43.7	11.3	0.20	0.12	0.08	0.06	13.54	3.42	SL
2	E	10-16	4.3	13.0	13.3	0.08	0.03	0.02	0.03	8.19	2.02	SL
3	BE	16-29	4.1	12.0	13.2	0.09	0.02	0.03	0.03	13.69	1.24	L
4	B1	29-71	4.7	9.8	5.3	0.09	0.05	0.06	0.02	10.68	2.08	CL
Pedon of Ta-Ta-Chia (TTC): 2620 m												
5	A	0-15	4.0	31.8	19.6	0.49	1.58	0.00	0.00	41.13	5.07	CL
6	E	15-27	3.8	9.0	16.6	0.39	1.51	0.00	0.00	29.31	6.49	SiL
7	BE	27-41	4.2	13.0	13.9	0.45	1.74	0.00	0.00	35.87	6.14	CL
8	Bw1	41-53	4.5	9.0	9.6	0.36	1.79	0.00	0.00	44.02	4.95	CL
Pedon of A-Li-Shan (ALS): 2350 m												
9	A	0-5	3.8	98.8	21.0	0.37	0.05	0.54	0.21	28.87	4.02	SiL
10	AE	5-11	3.7	31.8	9.5	0.17	0.04	0.11	0.06	13.04	2.86	SiL
11	E	11-25	4.1	5.8	11.1	0.07	0.04	0.01	0.03	9.13	1.68	L
12	Bhs1	25-50	4.2	11.8	14.9	0.10	0.03	0.02	0.03	19.14	0.96	L
Pedon of Shih-Tzu-Lu (STL): 1500 m												
13	A1	0-11	5.5	22.0	1.8	0.86	0.07	19.31	5.03	70.67	35.75	SL
14	A2	11-26	5.3	14.5	7.6	0.68	0.06	17.16	6.01	67.73	35.30	SL
15	AB	26-45	5.3	5.3	18.5	0.95	0.07	19.08	7.16	76.82	35.48	SL
16	B1	45-68	5.3	3.6	36.7	1.18	1.32	20.25	9.41	87.06	36.94	SL

[+]Texture: C=Clay, SCL=Sandy clay loam

Table 3. Free (Fed, Ald), noncrystalline (Feo, Alo) and organic-bound (Fep, Alp) sesquioxide contents (A) Lien-Hua-Chi and (B) A-Li-Shan regions

(A) Lien-Hua-Chi watershed region

Soil	Horizon	Free		Noncrystalline		Organic-bound	
		Fed	Ald	Feo	Alo	Fep	Alp
		---(g kg^{-1} soil)---					
Dystrochrept (Pedon 1)	A1	19.1	3.0	7.5	2.3	7.8	4.7
	A2	20.6	4.1	6.8	2.2	6.8	4.0
	Bw	22.2	4.2	4.3	2.1	4.7	3.2
	BC	21.2	4.1	2.5	2.0	3.6	2.6
Hapludult (Pedon 2)	A	6.6	2.1	2.3	1.5	3.7	2.2
	E	8.4	2.3	2.8	1.4	4.4	2.0
	E/B	9.0	2.5	2.0	1.4	4.4	2.0

Table 3. Continued

(B) A-Li-Shan region

Soil	Dithionite-citrate		Oxalate-oxalic acid		Sodium pyrophosphate			Extractable SO_4		Native SO_4
Pedon	Fe_d	Al_d	Fe_o	Al_o	Fe_p	Al_p	Fe_o/Fe_d	H_2O	KH_2PO_4	
	----------(g kg^{-1}) ----------							----------(g kg^{-1}) ----------		
Pedon of Lu-Lin-Shan (LLS): 2720 m										
1	2.93	2.32	2.06	1.77	0.51	1.47	0.71	0.031	0.348	0.379
2	18.84	2.90	2.17	1.55	1.82	1.26	0.12	0.052	0.026	0.078
3	27.12	8.05	10.96	4.06	13.32	3.36	0.40	0.190	0.378	0.568
4	27.91	6.36	11.31	4.60	18.67	5.37	0.41	0.024	0.073	0.097
Pedon of Ta-Ta-Chia (TTC): 2620 m										
5	2.76	0.60	0.98	0.44	0.99	0.43	0.36	0.002	0.065	0.067
6	2.55	0.47	0.65	0.30	0.75	0.27	0.26	0.008	0.069	0.077
7	4.73	0.93	0.84	0.42	1.46	0.48	0.18	0.020	0.117	0.137
8	4.72	0.96	0.76	0.39	1.32	0.44	0.16	0.313	0.130	0.443
Pedon of A-Li-Shan (ALS): 2350 m										
9	22.19	4.48	7.44	3.44	5.26	2.53	0.34	0.027	0.040	0.067
10	14.65	2.20	7.91	2.13	3.99	1.35	0.54	0.221	0.030	0.251
11	19.38	2.50	4.32	1.92	4.03	1.31	0.22	0.030	0.060	0.090
12	53.52	5.95	21.93	5.41	18.39	4.99	0.41	0.025	0.270	0.295
Pedon of Shih-Tzu-Lu (STL): 1500 m										
13	20.14	6.43	18.60	7.53	3.21	1.94	0.92	0.287	0.027	0.314
14	24.45	8.07	18.61	4.36	4.40	3.18	0.81	0.292	0.033	0.325
15	24.07	9.78	18.93	6.21	2.78	2.45	0.83	0.358	0.032	0.390
16	24.70	6.37	16.17	5.82	1.29	2.27	0.65	0.006	0.029	0.035

In general, Fed and Ald contents increased slightly with increasing soil depth, but Feo, Alo and Alp decreased with increasing soil pH. The sesquioxide and clay contents of pedon 1 were higher than tose of pedon in the Lieh-Hua-Chi region. (Table 3A). The Fed contents increased with depth except for the pedon of ALS (Table 3B). The Fed and Ald contents were lower in the Ta-Ta-Chia (TTC) pedon. Similarly, the Feo and Alo contents were the lowest in the TTC pedon. The Fep and Alp contents were less than Fed (Ald) or Feo (Alo) in all pedons and their distribution showed no significant trend with respect to soil depth.

4.2. Native Sulfate Sorption and Correlation with Soil Properties

The concentrations of water-soluble and native sorbed-SO_4^{2-} are presented in Table 4. The sorbed-SO_4^{2-} in the two pedons (Lien-Hua-Chi) extracted with NaH_2PO_4 in the ranged 0.14-1.09 mmol kg^{-1} was greater than that of native water-soluble sulfate. The original concentration of sorbed- SO_4^{2-} of pedon 1 was greater than that of pedon, due to pedon 1 contained more sesquioxides and clays than pedon 2. According to statistical analysis of sulfate sorption with respect to Fe and Al fractions, only exchangeable Al and Fep correlated highly with sorbed-SO_4^{2-} (p , 0.01 and $p < 0.05$). Before and after treatments with Na_2SO_4, exchangeable Al correlated highly with sorbed SO_4^{2-} ($r = 0.91$). The other soil properties showed no significant correlation with sulfate sulfate sorption. On the other hand, the Fed contents increased with depth except for the pedon of ALS in the Al-Li Shan region. The water-soluble and NaH_2PO_4-extractable sulfates showed no significant differences in all pedons. In the TTC pedon, the extractable sulfate increased with increasing soil depth, while it showed no such trends in the other pedons. Only water-soluble sulfate concentrations show high correlation with Fep ($r = 0.945$).

4.3. Effect of Reaction Duration and pH on Sulfate Sorption and Sorption Isotherm- Lien-Hua-Chi region

On the preliminary tests indicated that sulfate sorption was equilibrated completely within 1 h. Rajan (1978) [27] found that 95% of sulfate was sorbed on the sulface of Al-oxide within 1 h. The ^{32}S-isotope completed exchange with soil sulfur within a half hour [28]. Hence, the time for sulfate sorption to attain equilibrium was set at 2 h. The sulfate sorption data fitted poorly with the Langmuir equation. At high concentrations of Na_2SO_4, soil sorbed more sulfate with increasing sulfate concentrations, but at low Na_2SO_4 concentrations, soils desorbed the native sorbed- SO_4^{2-} to soil solutions. These results showed negative sorption; thus, no maximum monolayer sulfate sorption could be described from the Langmuir equation. The kinetic data of sulfate sorption can fit the pseudo-kinetic model well.

When tested with simulated acid rain, the effect of pH on sulfate sorption and released of cations was significant for soils treated with Na_2SO_4 solutions (0.1.mM) at pH 4.6 and 5.6. Soils from pedons1 and 2 showed desorption of native sorbed-SO_4^{2-} when treated with Na_2SO_4 solutions (0.1 mM) at pH 2.6. Released Al from soils showed the same trend of effect of pH as that for sulfate sorption (Figure 1). When the pH of soil solution was less than 4, the broken edge of clays and Al-oxides dissolved Al^{3+}, into soil solution and translocated in soil profiles [29].

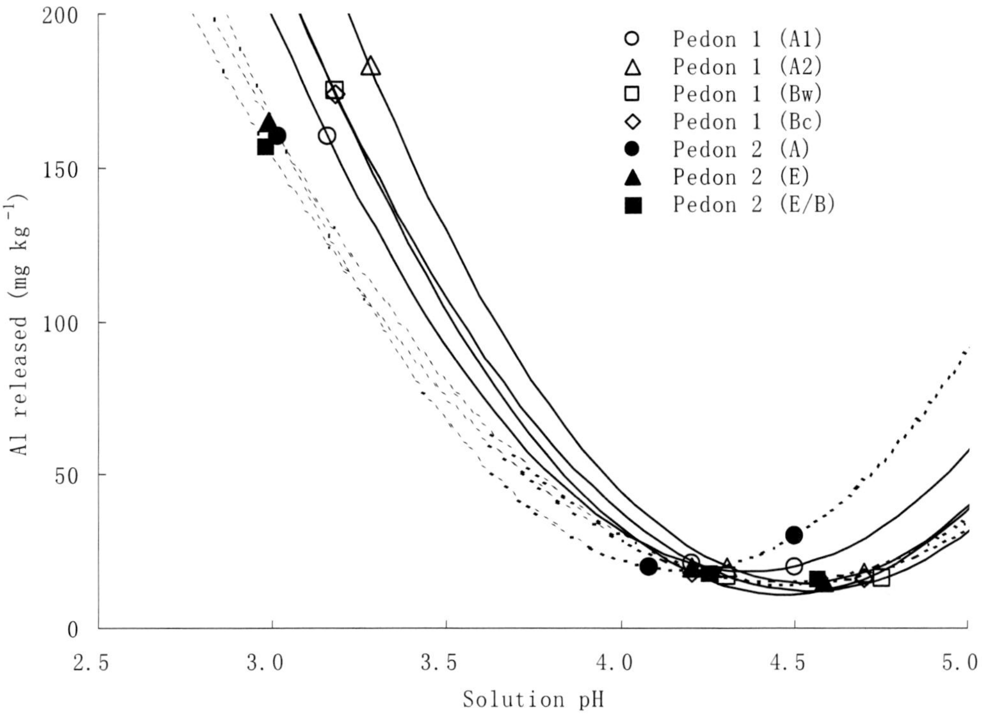

Figure 1. Aluminum released from soils treated with Na_2SO_4 solutions (0.1 mM) at varied pH.

4.4. Soil Columns Leached by Simulated Acid Rains

Introducing solutions of various acidity and sulfate concentrations into the leaching solution flowing through the soil column resulted in sharp increases in amount of sulfate in percolates after two-pore volumes in samples were replaced (i.e., pH 2.6 acid rain, pedon 1 (Bw horizon, 30-50 cm). The EC and the amount of Ca in percolates reached maximum values after 10-pore volumes in samples were replaced (Figure 2a and 2c). However, simulated acid rain of pH 3.6 and 4.6 showed no such significant differences. The concentration of Al, other cations and sulfate in percolates varied between soils (Figure 2d-2f).

Accumulation of leached cations of pedon 1 was greater than that of pedon 2. Cations retained in pedon 1 were also more than those in pedon 2. When these soils were treated with acid rain after 25-pore volumes in samples were replaced, the soils had diminished supplies of exchangeable cations. The residual exchangeable K, Na, Ca and Mg cations leached with simulated acid rain at pH 2.6 were less than those at pH 3.6 and 4.6 solutions. The exchangeable Al of acid rain at pH 2.6 was higher than at pH 3.6 and 4.6 (Table 4).

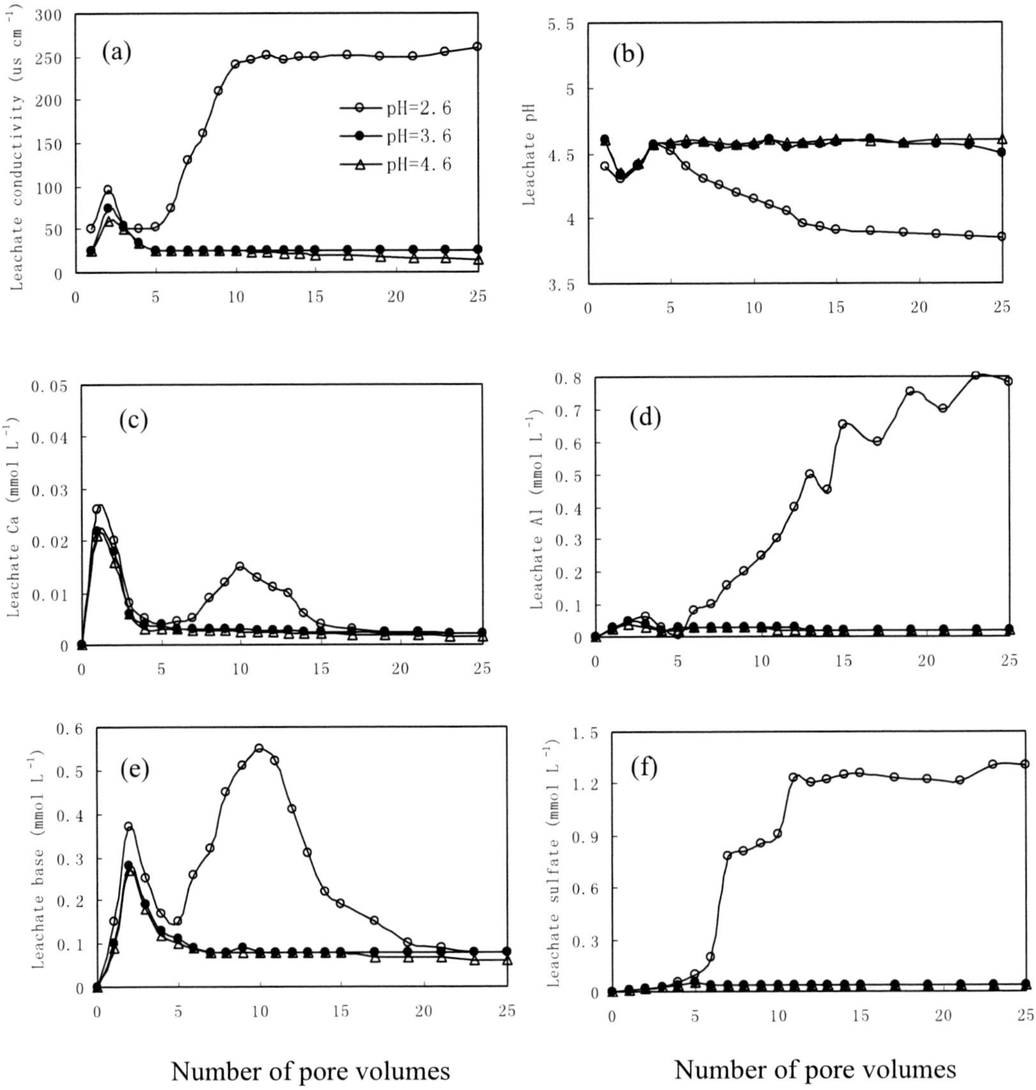

Figure 2. Variation of (a) EC, (b) pH, (c) Ca, (d) Al, (e) base cations, and (f) sulfate in percolates of soils leached with simulated acid rain after several pore volumes in soils were replaced (pedon 1, soil depth 30-50 cm)

Table 4. Comparison between original and released soil cations leached with simulated acid rain after 25-pore volumes in soils were replaced

Soil	Depth	Exchangeable cation				
		K	Na	Ca	Mg	Al
	(cm)	---mmol kg^{-1}---				
		---Initial---				
	0-10	2.98	0.05	0.11	0.68	5.50
Dystrochrept	10-30	2.29	0.06	0.04	0.37	5.94
(Pedon 1)	30-50	1.83	0.23	0.01	0.30	8.10
	50-80	1.95	0.29	0.03	0.21	9.16

Table 4. Continued

Soil	Depth	Exchangeable cation				
		K	Na	Ca	Mg	Al
	(cm)	------------------mmol kg^{-1}----------------------				
		----------------------Initial--------------------				
Hapludult (Pedon 2)	0-10	0.96	0.01	0.07	0.19	7.43
	10-40	0.55	0.11	0.04	0.09	5.83
	40-60	0.54	0.02	0.02	0.06	5.86
		--------------------- pH 2.6 of simulated acid rain 1--------------				
Dystrochrept (Pedon 1)	0-10	1.39	0.15	0.02	0.21	11.63
	10-30	1.28	0.17	0.01	0.05	17.55
	30-50	1.64	0.15	ND	0.04	21.80
	50-80	1.24	0.06	0.01	0.03	16.46
Hapludult (Pedon 2)	0-10	0.20	0.08	0.02	0.03	12.86
	10-40	0.23	0.10	0.02	0.02	10.97
	40-60	0.29	0.15	0.02	0.02	10.94
		----------------- pH 3.6 of simulated acid rain 2---------------				
Dystrochrept (Pedon 1)	0-10	1.85	0.07	0.20	0.83	9.10
	10-30	1.93	0.12	0.08	0.44	12.20
	30-50	1.86	0.23	0.02	0.36	6.24
	50-80	1.87	0.29	0.18	0.29	8.30
Hapludult (Pedon 2)	0-10	0.66	0.07	0.07	0.19	10.90
	10-40	0.44	0.05	0.01	0.03	9.41
	40-60	0.51	ND[a]	ND	0.03	3.81
		------------------ pH 4.6 of simulated acid rain 3--------------				
Dystrochrept (Pedon 1)	0-10	2.26	0.15	0.12	1.06	9.16
	10-30	1.20	0.12	0.06	0.48	11.62
	30-50	1.69	0.12	0.08	0.38	15.64
	50-80	1.78	0.11	0.08	0.28	19.13
Hapludult (Pedon 2)	0-10	1.05	0.05	0.05	0.23	11.91
	10-40	0.66	0.09	ND	0.08	9.44
	40-60	0.58	0.01	ND	0.06	8.82

[a] ND: not detected. Minimum detected limit (MDL= 0.1 mg kg^{-1}).
2. Aluminum released from soils treated with Na_2SO_4 solutions (0.1 mM) at varied pH.
3. Variation of (a) EC, (b) pH, (c) Ca, (d) Al, (e) base cations, and (f) sulfate in percolates of soils leached with simulated acid rain after several pore volumes in soils were replaced (pedon 1, soil depth 30-50 cm)

4.4. Soil pH Buffer Capacity by Titration Curve Method

According to the relationship between pH variation and acid titration, the pH buffer capacity could be divided into three linear phases (Table 5). The corresponding pH buffer capacity was named as pHBC-1, pHBC-2 and pHBC-3, respectively. pHBC-1 ranged from 0.98 to 2.56 cmol pH $unit^{-1}$ kg^{-1}; pHBC-2, from 5.42 to 17.50 cmol pH $unit^{-1}$ kg^{-1}; and pHBC-3, from 3.28 to 8.00 (Table 5). Obviously, the amount of pHBC-2 was greater than that of pHBC-1 and pHBC-3 in all soils. Among the pedons, only the values of pHBC-3 of TTC was higher than that of other pedons. However, no significant differences were present in other treatments and the pHBC variations of all pedons did not show any significant relationship with soil depth. According to the regression analysis of pHBC and soil properties, there existed significant correlations between pHBC-1 and clay content (r = -0.53*), Fep (r = 0.42*); pHBC-2 and pH (r = 0.65**), BS (r = -0.58**), Feo (r = -0.51*), Ald (r = 0.51*); pHBC-3 and Fed (r = -0.51*), Feo (r = -0.51*), Ald (r =0.51*), Alo (r = -0.59**), Alp (r = -0.44*), respectively. These results indicated that Fe- and Al-oxides had greater influence on soil pH buffer capacity [10].

Table 5. The pH buffer capacity of tested soils with three phases (cmol H^+ pH unit^{-1} kg^{-1})

Soil pedon	pHBC-1	pHBC-2	pHBC-3
Lu-Lin-Shan (LLS)			
A	0.98	10.48	3.28
E	1.18	10.07	3.45
BE	2.11	8.83	3.57
B1	2.00	7.53	3.51
Ta-Ta-Chia (TTC)			
A	1.85	17.50	6.90
E	1.87	12.44	6.86
BE	1.80	14.00	6.90
Bw1	2.44	12.39	8.00
A-Li-Shan (ALS)			
A	2.15	10.57	3.51
AE	1.27	10.11	3.57
E	1.53	10.07	3.51
Bhs1	2.56	9.12	3.70
Shih-Tzu-Lu (STL)			
A1	1.54	5.42	3.92
A2	1.65	6.39	3.85
AB	2.02	6.53	3.77
B1	1.87	7.14	3.85

4.5. Theoretical Calculations of Proton Affinity of Study Soils

Two assumptions were used when calculating the proton consumption of the studied soils (Table 6). The first assumption was that the ionic activity coefficient (Υ_i) during the titration period was equal to 1, and the second was that the ionic activity coefficient remained constant from the initial to final stage, the initial ionic activity coefficient was calculated using Davies' equation. The slope values obtained from the two assumptions were different but it showed a similar trend. The proton affinity potentials increased with increasing soil depth except for the STL pedon. Meanwhile, the other three pedons showed a similar range of proton affinity potentials between 2-20 cmolkg^{-1} mM^{-1}, but the STL pedon (i.e., 1500 m in elevation) had much higher values than the other pedons. Further statistical analysis showed that the soil proton affinity potential was correlated with soil pH (r = 0.90**), CEC (r = 0.91**), BS% (r = 0.97**), and X-ray noncrystalline sesuioxides (Feo, r = 0.68**), respectively [10].

Table 6. Proton affinity of the study soils

Soil. pedon	Horizon	Depth (cm)	γi=1		γi constant	
			slope	R^2	slope	R^2
Pedon Lu-Lin-Shan (LLS)						
1	A	0-10	2.757	0.880	2.169	0.860
2	E	10-16	2.419	0.879	1.880	0.857
3	BE	16-29	6.406	0.941	5.296	0.932
4	B1	29-71	10.62	0.940	8.922	0.932
Pedon Ta-Ta-Chia (TTC)						
5	A	0-15	14.52	0.902	12.24	0.909
6	E	15-27	7.974	0.975	6.616	0.972
7	BE	27-41	12.53	0.876	10.63	0.883
8	Bw1	41-53	18.28	0.873	15.37	0.874
Pedon A-Li-Shan (ALS)						
9	A	0-5	6.449	0.951	5.345	0.949
10	AE	5-11	2.999	0.933	2.370	0.922
11	E	11-25	2.424	0.899	1.888	0.882
12	Bhs1	25-50	12.27	0.955	10.32	0.951
Pedon Shih-Tzu-Lu (STL)						
13	A1	0-11	86.74	0.902	74.53	0.895
14	A2	11-26	89.52	0.944	77.05	0.940
15	AB	26-45	96.65	0.926	82.94	0.920
16	B1	45-68	84.21	0.943	72.18	0.937

Conclusions

The comparison of native sorbed SO_4^{2-} and the maximum monolayer sulfate sorption indicated that forest soils in the Lieh-Hua-Chi watershed were almost saturated with SO_4^{2-} anions. Sulfate sorption fitted poorly to the Langmuir equation at low concentrations of Na_2SO_4. High clay and sesquioxide contents of pedon 1 acted as a sink to retain sulfate when soil columns were leached with simulated acid rains. Pedon 2 had low exchangeable cations after 25-pore volumes in soil samples were replaced. These forest soils in the Lien-Hua-Chi watershed area showed a H^+ buffer capacity. Pedon 1 showed a better buffer capacity of acid, greater amounts of cations and more sulfate retention than pedon 2. Pedon 1 was more resistant to the impact of acid rain than pedon 2.

Higher soil pH, CEC and BS were present in the lower elevation pedon of the A-Li Shan region, suggesting that soil potential pH buffer capacity decreased with increasing elevation. The proposed three-phase pHBC could well describe the pH titration curves, but pHBCs showed no significant differences with soil pedons and horizons. The theoretical calculation of proton affinity showed much higher buffering potentials in the lower elevation site and well correlated with soil pH, CEC, BS and Feo contents. Thus, we suggest that it is important to consider the pH buffer capacity together proton consumption capacity when evaluating soil-buffered acidification.

Acknowledgments

We thank the National Science Council, Taiwan, Republic of China, for financial support under the projects of forest soils (NSC #85-2621-B002-017; 86-2621-B002-022-A07, 87-2621-B002-053-A10, and 96-2621-B002-007).

References

[1] King HB, Shiue ML: Acidic precipitation and nutrient cycling of a subtropical montane hardwood watershed in central Taiwan, *Int. Symp. on Air Poll. Soil Micro. and Biotech. Forests*, June 1-5, 1992, Chitou, Taiwan. 1992.

[2] Chen JH, Chang KH, Jeng FT: *The acid rain in Taiwan area. Int. Conf. on Reg. Environ. and Clim. Chain in East Asia,* Nov. 30-Dec. 3, Taipei, pp. 28-31. 1993.

[3] Howarth RW, Stewart JWB, Ivanov MV: Sulphur Cycling on the Continents: Wetlands, Terrestrial Ecosystems and Associated Water Bodies. *Scope* **48**. John Wiley and Sons. New York. pp. 350. 1988.

[4] Rechcigl JE, Sparks DL: Effect of acid rain on soil environment: A review. *Commun. Soil Sci. Plant Analy.,* **16**(7):653-680. 1985.

[5] Sanger, L.J., Billett, M.F. and Cresser, M.S., 1994. Changes in sulphate retention, soil chemistry and drainage water quality along and upland soil transect. *Environ. Poll.,* **86**(2):119-128.

[6] van Breemen N: Soil forming processes in acid sulfate soils. In H. Dost (Editor) *Acid Sulphate Soils.* Proc. Int. Symp. 12-20, Aug., Washington, Vol. 1. International Institute for Land Reclamation and Improvement, Wageningen, Netherlands, pp. 66-130. 1972.

[7] Johnson DW, Todd DE: Relationships among iron, aluminum, carbon and sulfate in a variety of forest soils. *Soil Sci. Soc. Am. J.*, **47**: 792-800. 1983.

[8] Ulrich B: The production and consumption of hydrogen ions in the ecosphere. In: T.C. Hutchinson and H. Havas (Editors.). *Effects of Acid Precipitation on Terrestrial Ecosystem.* Plenum Press, New York, pp. 255-282. 1980.

[9] King HB, Wang MK, Zhuang SY, Hwong JL, Liu CP, Kang MJ: Sorption of sulfate and retention of cations in forest soils of Lien-Hua-Chi watershed in central Taiwan. *Geoderma* **143**:153. 2006.

[10] Zhuang SY, Wang MK, King HB, Hwong JL, Hsu FS: Rain acid buffer capacities of alpine forest soils in central Taiwan. *Geoderma* **137**:174-178. 2006.

[11] Clayton JL, Kennedy DA, Nagel T: Soil response to acid deposition, Wind River mountains, Wyoming: I. Soil properties. *Soil Sci. Soc. Am. Proc.* **28**: 632-635. 1991.

[12] Parfitt RL, Smart RSC: The mechanism of sulfate adsorption on iron oxides. *Soil Sci. Soc. Am. J.* **42**: 48-50. 1978.

[13] Harrison RB, Johnson DW, Todd DE: Sulfate adsorption and desorption reversibility in a variety of forest soils. I. *J. Environ. Qual.* **18**: 419-426. 1989.

[14] Parks GA: The isoelectric points of solid oxides, solid hydroxides and aqueous hydroxo complex systems. *Chem. Rev.* **65**:177-198. 1965.

[15] Hsia YJ, Liu LY: Optimum water year selection for small forest watersheds. *Taiwan For. Res. Ins. Bull.* **353**. Taipei. 1981.

[16] Liu V: The forest vegetation of Taiwan. *Quar. J. Chinese For. Sci.* **5**:56-86. 1972.

[17] Ho CS: *An Introduction to the Geology of Taiwan: Explanatory Text of the Geological Map of Taiwan.* Publ. by the Ministry of Economic Affairs, Taipei, Taiwan, pp. 153. 1975.

[18] King HB, Torrance JK, Bowen LH, Wang C: Iron concretions in a Typic Dystrochrept in Taiwan. *Soil Sci. Soc. Am. J.* **54**: 462-468. 1990.

[19] Soil Survey Staff: *Keys to Soil Survey Taxonomy.* U.S. Dept. of Agriculture, Soil Management Support Services. 17th ed. US Government Printing Office, Washington, DC. 1998.

[20] Walkley A, Black CA: In experimentation of the Detjareff method and a proposed modification of the chromic acid titration method. *Soil Sc*i. **85**: 307-318. 1934.

[21] Jackson ML:.*Soil Chemical Analysis.* Advanced Course, 2nd ed., University of Wisconsin, Madison, WI. 1979.

[22] Bouyoucos GJ: Hydrometer method improved for making particle-size analyses of soils. *Agron. J.* **54**: 464-465. 1962.

[24] Loveland PT, Dighy P: The extraction of Fe and Al by 0.1 M pyrophosphate solutions: comparison of some techniques. *J. Soil Sci.* **35**: 243-250. 1984.

[23] Mehra OP, Jackson ML: Iron oxides removal from soils and clays by a dithionite-citrate system, buffered with sodium bicarbonate. *Clays Clay Miner.* **7**: 317-327. 1960.

[25] Helayer KR, Porter WM: 'Soil acidification, its measurement and the processes involved', In: *Soil Acidity and Plant Production.* A.D. Robson (editor). Academic Press, Sydney, Australia, pp. 61-101. 1989.

[26] Contescu C, Jagiello J, Schwarz JA: Heterogeneity of proton binding sites at the oxide/solution interface. *Langmuir* **9**:1754-1765. 1993.

[27] Rajan SSS: Sulfate adsorbed on hydrous alumina, ligands displaced, and changes in surface charge. *Soil Sci Soc Am. J.* **42**:39-44. 1978.

[28] Singh BR: Sulfate sorption by acid forest soils: 2. Sulfate adsorption isotherms with and without organic matter and oxides of aluminum and iron. *Soil Sci.* **138**:

[29] Mason B*: Principle of Goechemistry.* Wiley, New York. pp. 167. 1980.

INDEX

B

C

D

E

F

G

H

I

J

K

L

N

O

P

Q

R

S

T

U

V

W

X

Y